Make:

Getting Started
with the micro:bit

Coding and Making with the BBC's
Open Development Board

Wolfram Donat

MAKER MEDIA

SAN FRANCISCO, CA

Printed in the United States of America.

Published by Maker Media, Inc., 1700 Montgomery Street, Suite 240, San Francisco, CA 94111

Maker Media books may be purchased for educational, business, or sales promotional use. Online editions are also available for most titles (safaribooksonline.com). For more information, contact our corporate/institutional sales department: 800-998-9938 or corporate@oreilly.com.

Publisher: Roger Stewart
Editor: Patrick DiJusto
Copy Editor and Proofreader: Elizabeth Welch, Happenstance Type-O-Rama
Interior Designer and Compositor: Maureen Forys, Happenstance Type-O-Rama
Cover Designer: Maureen Forys, Happenstance Type-O-Rama
Indexer: Valerie Perry, Happenstance Type-O-Rama

August 2017: First Edition

Revision History for the First Edition
2017-08-08 First Release

See oreilly.com/catalog/errata.csp?isbn=9781680453027 for release details.

978-1-680-45302-7

Safari® Books Online

Safari Books Online is an on-demand digital library that delivers expert content in both book and video form from the world's leading authors in technology and business. Technology professionals, software developers, web designers, and business and creative professionals use Safari Books Online as their primary resource for research, problem solving, learning, and certification training. Safari Books Online offers a range of plans and pricing for enterprise, government, education, and individuals. Members have access to thousands of books, training videos, and prepublication manuscripts in one fully searchable database from publishers like O'Reilly Media, Prentice Hall Professional, Addison-Wesley Professional, Microsoft Press, Sams, Que, Peachpit Press, Focal Press, Cisco Press, John Wiley & Sons, Syngress, Morgan Kaufmann, IBM Redbooks, Packt, Adobe Press, FT Press, Apress, Manning, New Riders, McGraw-Hill, Jones & Bartlett, Course Technology, and hundreds more. For more information about Safari Books Online, please visit us online.

How to Contact Us

Please address comments and questions to the publisher:

Maker Media
1700 Montgomery St.
Suite 240
San Francisco, CA 94111

You can send comments and questions to us by email at books@makermedia.com.

Maker Media unites, inspires, informs, and entertains a growing community of resourceful people who undertake amazing projects in their backyards, basements, and garages. Maker Media celebrates your right to tweak, hack, and bend any Technology to your will. The Maker Media audience continues to be a growing culture and community that believes in bettering ourselves, our environment, our educational system—our entire world. This is much more than an audience, it's a worldwide movement that Maker Media is leading. We call it the Maker Movement.

To learn more about Make: visit us at makezine.com. You can learn more about the company at the following websites:

Maker Media: makermedia.com
Maker Faire: makerfaire.com
Maker Shed: makershed.com

This book is dedicated to Becky and Reed, who put up with a husband and father who disappears into the workshop or office for extended periods of time on a fairly regular basis. On a related note, they also deal with a number of screwy creations flowing out of the aforementioned workshop.

Contents

Acknowledgments

Patrick is an awesome editor who not only makes sure that I mean what I say and I say what I mean, but is also fun to work with. Bob checks that stuff works the way I say it does. Liz catches all the last-minute errors and makes sure that the end product of our collaboration looks as great as it does.

And last but not least, Oliver helps to make sure the office door remains in working order, Chloe ensures that all mobile creations are capable of evasive maneuvers, and Smudge both gives and receives emotional support.

I couldn't do it without you guys.

About the Author

Wolfram Donat is an engineer, a writer, and a maker who has written books on subjects ranging from home-built animatronics to Windows XP to using the Raspberry Pi in your projects. His varied interests include robotics, embedded systems, autonomous underwater vehicles, computer vision, and the Internet of Things.

He received his degree in computer engineering from the University of Alaska, Anchorage and has received funding from NASA for work on autonomous submersibles. He currently lives in Southern California with his wife, son, and a small menagerie.

Introduction to the micro:bit

If you've been paying attention to news in the world of technology, you may have noticed that there seems to be an astonishing number of single-board computers (SBCs) hitting the market lately. In a wave of devices that may have started with the Raspberry Pi, there are now dozens of small, powerful devices, ranging in price from a few dollars to a few hundred dollars. The Pi Zero, the Raspberry Pi Foundation's lowest-cost board, uses a small, 1 GHz, single-core ARM chip and costs about five dollars. On the other end of the spectrum, the NVIDIA Jetson TK1 includes both an ARM A57 quad-core chip *and* a 256-core Maxwell GPU and will cost you about six hundred dollars. It is, however, still considered a single-board computer. Kickstarter is full of new SBCs, some successful, some not.

At the same time as the release of all of these surprisingly powerful small computers, various technology companies have been quietly releasing a flood of even smaller, lower-power chips and devices in the background. These boards are powered by a variety of processors, from ARM CPUs to smaller microcontrollers like the Atmega 328, and they are usually designed mainly for one purpose: performing one or more simple tasks and then interfacing with the Internet of Things (IoT).

What is the Internet of Things? For the full story behind the IoT, check out the accompanying sidebar. The short version is that the IoT is a worldwide web of small, low-power devices that are able to communicate with other devices—both IoT devices and more full-featured machines such as smartphones and computers—via the Internet and other smaller networks. These devices are meant to connect everything, from your home thermostat to your refrigerator to your toaster to your keychain, and allow them to communicate via a network. They must necessarily subsist on almost no power (there's no room for big, bulky batteries in your keychain) and thus must also be sort of stupid, CPU-wise. The vast majority of them don't need to be particularly powerful, though; many times their main function is simply to collect data and relay it to a more powerful computer, smartphone, or tablet, or perform a simple task in response to a simple command from another device.

What is the IoT?

The Internet of Things, or "IoT" for those in the know, is what many call a natural evolution of technological advances in shrinking microprocessors, low-power devices, and networking connectivity. It's pretty much accepted that it's the next step in wiring the planet together and making all of our stuff smarter.

Up until a decade ago or so, computer processors were still too big and required too much power to be used in anything other than servers, desktops, laptops, and maybe a few clunky tablets. Firms like Intel were still trying to shrink the chips, not really worrying about putting them into small devices.

The huge growth of the cell phone changed that, however. It is projected that by 2020, seventy percent of the world's population over the age of six will have a mobile phone and 90 percent will be covered by broadband networks. The upswing in mobile phone usage, and the associated need for smaller, faster, more powerful devices, sparked a new arms race in small, low-power chips, which in turn were used in devices other than smartphones, like the Raspberry Pi and the micro:bit. Developers and companies began to realize that they could use these chips to connect pretty much anything over a network, and the IoT was born.

The IoT is still in its infancy, despite experts continuing to tout it as the next big thing (guilty!). It is starting to take hold, though, as evidenced by devices such as the Nest thermostat, the Ring doorbell, and now the Google and Amazon digital home assistants. Internet connectivity is beginning to show up in smart refrigerators, and it's only a matter of time before your washing machine will be able to text you when your load of delicates is done. For that is one of the goals of the IoT: to allow things that you interact with on a daily basis to have enough brains to make your life easier.

These small IoT development boards also run the gamut of prices but generally stay below the one hundred dollar mark. The C.H.I.P. is the newest kid on the block (Figure 1.1) and costs only about nine dollars. The Particle Photon has an ARM Cortex processor and will cost you about twenty dollars, and the Intel Edison—the powerhouse of the group—costs about seventy dollars (Figure 1.2).

FIGURE 1.1: The C.H.I.P.

FIGURE 1.2: Intel Edison

Into this group we have a newcomer: the BBC micro:bit. This little device is being given away, at no cost, to a million 11-year-old students in the United Kingdom, courtesy of the British Broadcasting Corporation's Make It Digital initiative. This initiative is part of a push to increase digital skills among British youngsters, as—according to the BBC—there is a significant digital skills shortage in the country, with 1.4 million skilled professionals projected to be needed in the next five years. The BBC is following in the footsteps of the Raspberry Pi Foundation; organizers of the initiative believe that by giving young people access to cheap or free computers and technology, they can stimulate a lifelong interest in learning, programming, and engineering. The company is partnering with about 30 other organizations to bring this idea to fruition. Some are offering financial assistance; others, like element14, are helping manufacture the boards, and so on. It's truly a group effort on the part of many different technological associations in order to get kids interested in technology.

The micro:bit itself is a tiny device (Figure 1.3)—4 by 5 centimeters to a side—and can be powered by a variety of sources: two AAA batteries, a coin cell, a USB connection, or any other source of 3 volts. The board connects to your Windows, Mac, or Linux computer with a standard micro USB cable and mounts as an external drive or device without any driver software necessary. It is equipped with twenty-five LEDs and two small buttons, all of which are programmable, that allow the user to interact with it. The bottom of the board is lined with twenty general-purpose input/output (GPIO) pins, which are accessible with either alligator or banana clips, or by inserting the board into a special connector that connects headers to all of these pins. The micro:bit also has an onboard compass and accelerometer that you can read from and incorporate into your scripts, and it has an onboard Bluetooth Low-Energy (BLE) antenna that enables it to pair with any Bluetooth-enabled devices, such as your smartphone or laptop.

FIGURE 1.3: The BBC micro:bit

In the interest of keeping things simple so that kids stay interested and involved, the Micro:bit Foundation offers five (as of this writing) different ways to program the micro:bit. All five are web-based programming environments for different languages, written in JavaScript. Students can choose Touch Develop by Microsoft, Code Kingdoms' JavaScript editor, Microsoft's Block Editor (Figure 1.4), MicroPython (Figure 1.5), or PXT, again by Microsoft. All of these environments vary according to skill level; some are simple block-based programming similar to Scratch (students move code-based puzzle pieces into place in a graphical environment) whereas others are fully text based for more advanced coders. There are other ways to program the onboard processor that go beyond these simple web-based environments, which we'll get into later.

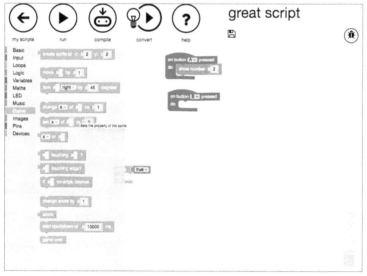

FIGURE 1.4: Microsoft Block Editor environment

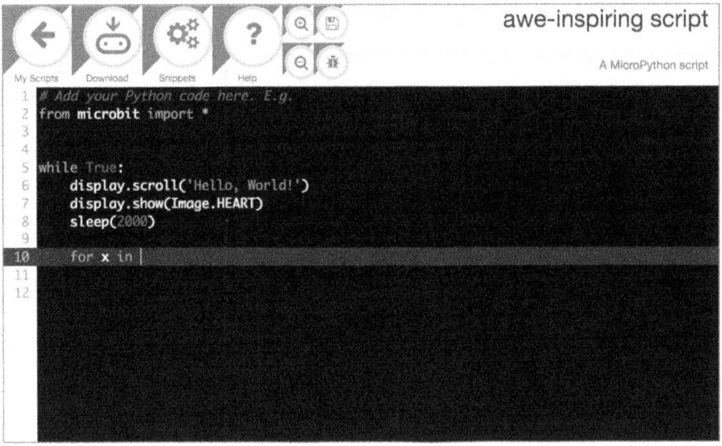

```
1  # Add your Python code here. E.g.
2  from microbit import *
3
4
5  while True:
6      display.scroll('Hello, World!')
7      display.show(Image.HEART)
8      sleep(2000)
9
10     for x in |
11
12
```

FIGURE 1.5: **MicroPython environment**

So how do we go about procuring one of these awe-some boards? As I write this, the micro:bit is now available all around the world, though it was originally only available from UK-based sellers. If you prefer to order from overseas, however, https://ww.kitronik.co.uk has always been a reliable source, and that's where I've gotten all of my micro:bits and the associated accessories.

Speaking of accessories, I highly recommend doing some more serious shopping when you buy your board, because you're more than likely going to want to do more with your board than just program a few simple scripts and games. When you go online to purchase your micro:bit, do a search for "micro:bit accessories" and see what's available. Here's what I bought to go along with mine:

- The Edge Connector Breakout Board (Figure 1.6)— This is almost a must-have, because without it you will have an extremely difficult time accessing the

GPIO pins on the bottom of the board. Once you slide the micro:bit into the slot on the edge connector, all of its pins are mapped to a row of headers, allowing you to use the jumper wires you probably already have in your toolkit.

FIGURE 1.6: The Edge Connector Breakout Board

- The Edge Connector Motor Driver Board (Figure 1.7)— If you want to use your micro:bit to do anything really cool, like, say, drive a robot in response to signals from your cell phone, you're going to want this board. Again, you slide the micro:bit into the slot, and all the pins you need to connect to external motors and power sources are instantly made available to you. It's very handy and will save you a lot of time when it comes to building.

FIGURE 1.7: The Edge Connector Motor Driver Board

☑ The MI:power board (Figure 1.8)—I don't consider this board as vital as the other two I've mentioned, but it's definitely useful nonetheless. First, it allows you to power your micro:bit using a coin cell battery rather than a bulky pair of AAAs. Second, it mounts to your board as a shield, similar to the Arduino shields, making everything a one-piece set. And finally, it has a piezoelectric buzzer built in, letting you experiment with the sound-making capabilities of the micro:bit without having to make any extra connections.

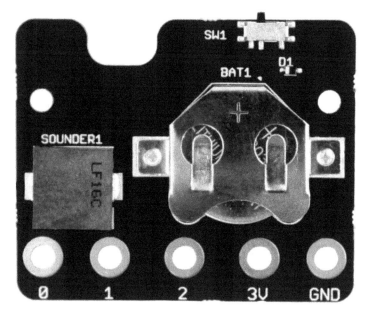

FIGURE 1.8: The MI:power board

I'm going to proceed with this introductory chapter on the assumption that you have at least the basic micro:bit in your hot little hands and are itching to get to do something with it. We'll get to using the accessories later.

For now, let's jump in, power on the micro:bit, and see what it does. Connect it to an available USB port on your computer using a micro USB cable (your device probably came with one; if it didn't, any standard micro USB cable will work). The board LEDs will flash in a square pattern, and then "HELLO" will scroll across the board (Figure 1.9). If by chance your board doesn't have this introductory program loaded onto it, you can download it here: https://github.com/wdonat/microbit-code/blob/master/chapter1/MicroBit-First-Experience.hex.

FIGURE 1.9: "HELLO" on the micro:bit

Next, the LEDs will flash "A," followed by an arrow pointing to the button on the left. When you push the button, the LEDs flash another pattern, and then "B" and a right-pointing arrow. Pushing this button results in another pattern, and then the message "SHAKE!" Shaking the board results in yet another pattern, followed by the message "CHASE THE DOT." This is a little game that allows you to tilt the micro:bit to make one light "follow" the one lit by the device. Once you "catch" it once or twice, the board replies with "GREAT! NOW GET CODING!" and a little heart symbol that flashes on and off (Figure 1.10).

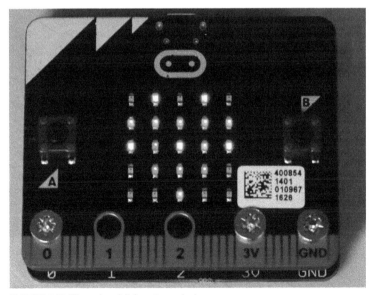

FIGURE 1.10: The micro:bit heart symbol

This little introductory display may not seem like much, but it's an enlightening intro into some of what the board can do. The LEDs can be lit in any pattern you like, including scrolling text—this addresses the problem of interacting with the user without a screen or monitor. (In fact, later on when you're programming, any error messages actually scroll across the LED array. Handy, if a bit difficult to read.) You can program the two buttons to respond to presses, and you can also use the onboard accelerometer to respond to user inputs such as shaking. In fact, about the only things on the board *not* introduced by the introductory program are the onboard compass/magnetometer and the Bluetooth Low Energy antenna (BLE). Don't worry—we'll get to those soon enough.

Now that you've seen the default program, let's go through the process of putting our own simple program onto the board. Probably the easiest way to program the micro:bit is while it's still attached to the computer with the USB cable. If you open your Windows Explorer window or the Mac Finder window, you should see the micro:bit showing up as an external drive or device (Figure 1.11.)

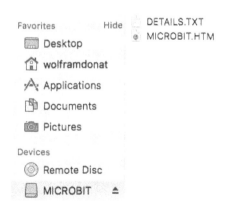

FIGURE 1.11: micro:bit Finder window (Mac)

When you write programs (also called *scripts*) and download them to your computer, you'll move the resulting hex file to the MICROBIT drive you see here. This will flash the device with your program. You can also flash the device with a Bluetooth connection, but we'll get to that later.

For now, let's experiment with adding another program. Open a browser window and visit https://github.com/wdonat/microbit-code/tree/master/chapter1. Right-click on the `microbit-astounding-script.hex` file and save it to your machine. When the file is downloaded (it shouldn't take long—it's only 584 KB), drag the script onto your MICROBIT device and let it do its thing. The transfer should only take a few seconds. When the device has finished flashing (signified by the yellow LED ceasing to flicker), your computer may complain that you should properly eject the MICROBIT device; ignore it and just close the warning pop-up window.

This is a simple script; it just counts the number of times you press the A button (on the left) and displays that count after each press, up to 150 presses. The actual script is seven lines of code, though the hex file is much more—about 13,000 lines. (If you're interested, right-click on the downloaded hex file and open it with a text editor like Notepad or Sublime Text, and you can see what the script looks like after it's been compiled. Definitely *not* user-friendly.) The script is straight from the Microsoft Touch Develop page for the micro:bit (one of the programming editors that are available to use) and looks like this:

```
function main()
    var counter := 0
    input → on button pressed (A) do
        counter := counter + 1
        basic → show number (counter, 150)
    end
end function
```

I won't go too deep into detail about this script now, because it's highly language-dependent and I don't plan to cover Touch Develop until one of the appendices, but a quick look over the code shows how basic the commands really are. You declare a variable (counter), look for some inputs, tell the board what to do with those inputs, and finally end the function. It really doesn't get much simpler than that.

That is a very high-level introduction to the micro:bit, what it can do, and how to interact with it. In the next chapter, I'll go over what exact hardware is on the board, as well as the add-ons we talked about.

A Tour of the micro:bit

Now that you've had a quick introduction to what the micro:bit is and what it can do, it's time to take a look at this impressive little device and see what's packed onto its small surface (4.5 × 5 cm—it's been billed by the BBC as being about half the size of a credit card). I usually introduce new users to a device like this by examining each component one by one, moving clockwise around the board, and that seems like a perfectly reasonable route to take now. I'll refer to the side of the micro:bit with the USB power connector and the micro:bit logo as the *back*, and the side with the array of LEDs and the two push buttons as the *front* (Figures 2.1 and 2.2).

Starting at the top (12 o'clock position) on the back, we have a standard USB micro port (*not* a USB *mini* port). When you connect the micro:bit to your computer, the port is used both to power the board and for data transfer from your computer. The board requires 3.3V to operate. USB offers around 5V, so a regulator is used to lower the input voltage to a level that the board can use when it's being powered by your computer.

FIGURE 2.1: Back of the micro:bit

FIGURE 2.2: Front of the micro:bit

Keep in mind, however, that this port is not necessary for either power or data transfer. You can power the micro:bit with a battery pack, and you can load programs to the board (also called *flashing* the board) by way of Bluetooth and a Bluetooth-enabled device such as a smartphone or laptop. This can come in handy if you install your micro:bit in an inaccessible place such as a project box or buried deep inside a load of electronics and wiring; you can flash the board with a new program by merely coming within a few feet of it. We'll go over Bluetooth connectivity and flashing in Chapter 3, "Programming Using MicroPython," and Chapter 7, "Bluetooth."

Just next to the USB port, before the push button, is a small yellow LED that you probably wouldn't notice until you plug the unit into your computer. It's a status LED, and its purpose is simply to let the user know that the micro:bit is doing something, whether it's loading a program or sending data.

Next to the USB port and the status LED is a momentary push button that serves as the reset button. When the board has a program onboard and is executing it, pushing this button resets it to the start of that program, as if the board had been powered off and then on again. This button is not programmable by the user; it's hard-coded as a reset button only. If you've played around with an Arduino, you're familiar with the concept of this button and what it's used for. It's helpful if your board freezes, if you need to restart a program for any reason, or if you just need to reset the board to a last-known-good configuration.

Next to the reset button is another power port. This port has two pins and is where you'll plug in an external power source if you're not powering the device via USB. The basic device comes with a battery pack that holds two AAA batteries; the Molex female connector to that pack plugs into the male pins on this connector.

Continuing in a clockwise direction, you'll see a small black integrated circuit (IC) set back a bit from the edge of the board (Figure 2.3). This is the USB controller that allows the CPU to communicate with the USB port. It's an ARM Cortex-M0+ chip that not only allows the USB communication to take place, it also regulates the 5V power from the USB port down to the 3.3V, which the micro:bit needs to operate. The regulator portion of the chip isn't necessary or used if you're powering your board with batteries.

FIGURE 2.3: The USB microcontroller

Now we come to the bottom of the board and its piano-key-like appearance. Each of these twenty-five individual metal-plated "stripes" is a general-purpose input/output (GPIO) pin, which can be accessed by the user. It can be a difficult thing to do if you don't have an edge connector (see Chapter 1, "Introduction to the micro:bit"), but the pins labeled 0, 1, 2, 3V, and GND (on the front of the board) are easily accessible with either a small alligator clip or a banana plug (Figure 2.4).

FIGURE 2.4: **Simple connections**

Still moving clockwise, directly above the pins on the left is the first of the onboard sensors, the accelerometer. This miniscule black IC is a Freescale MMA8652 full-fledged three-axis accelerometer that communicates with the processor using the I²C protocol. It has 12 bits of resolution and communicates with data rates from 1.56 Hz to 800 Hz— quite a wide range of possibilities, depending on your needs and your project. No, it's not a professional nine-axis inertial measurement unit (IMU) like you will find in many drone autopilots (for example), but three axes should be plenty for most simple micro:bit projects. You always have the option of upgrading and connecting a more powerful sensor via the GPIO pins, should your project call for it.

Next to the accelerometer is the other onboard sensor, the compass/magnetometer. Similar to the accelerometer, this IC is a Freescale MAG3110 three-axis digital magnetometer. It can be used as either a compass or a metal detector, and like the accelerometer communicates with the CPU over the I²C bus. It measures magnetic fields with an output data rate of up to 80 Hz, and has a sensitivity of 0.1 microteslas. Like the accelerometer, we'll explore how best to communicate with it in a later chapter.

The I²C Protocol

I²C (or I squared C or I-I-C) stands for inter-integrated circuit, and is a communications protocol that was developed by Philips Semiconductor and released back in 1982. It's a multi-master, multislave protocol that allows multiple devices to communicate with each other over typically short distances. I²C is a serial bus that's often used with microcontrollers, sensors (like those on the micro:bit board), and small embedded devices. Most single-board computers like the Raspberry Pi and many, many sensors, ranging from barometers to GPS modules to magnetometers to thermometers and others, have native support for the I²C protocol, and it remains one of the easiest, most basic ways to communicate with external devices and sensors from a central CPU.

After these two sensors we come to the heart and brain of the whole thing—the processor (Figure 2.5). This little black square is a 32-bit ARM Cortex M0 processor with 256 KB of flash memory and 16 KB of RAM, running at 16 MHz. It's Bluetooth-capable, with an embedded 2.4 GHz Bluetooth low-energy transceiver.

So what does all of that mean in the context of capabilities and power? First of all, it's a 32-bit machine, so it's not quite as fast or powerful as the 64-bit processors we're all getting used to. However, it's more than fast enough for a tiny machine like this. The 256 KB of flash memory refers to the memory that is retained when there is no power; in other words, when you unplug the micro:bit from your computer or from its battery pack, the contents of flash memory are retained, sort of like the hard drive on your computer or laptop. This is where your hex files are stored and is why the program will repeat every time you power on the device. Now, 256 KB may not seem like a lot of memory (most JPEG files are bigger than that, for example), but the hex files your programs are stored in are tiny. A 256 KB hex file would be quite a hefty program.

The contents of the 16 KB of RAM, on the other hand, disappear every time the device loses power, just like the RAM in your computer. This batch of memory is where the micro:bit performs calculations; it moves data from the registers into RAM, does what it needs to, and then moves it out again. Because 16 KB is not a whole lot of space, it limits the micro:bit's capabilities, but the board was never designed to do a lot of heavy lifting, compute-wise. Instead, it makes more sense to farm calculations and computations out to another, more powerful device, such as a smartphone, and merely use the micro:bit to collect and display data. It's helpful to remember that the micro:bit, like other IoT platforms, is necessarily a very low-power device, and a more powerful onboard CPU would use unhealthy amounts of power. It's quite impressive that the ARM chip is as powerful as it is for the amount of power it uses—at most around 0.03 watts, or about a one-hundredth as much as a standard night light.

Finally, to complete our journey around the back side of the board, we come to the almost invisible Bluetooth Low-Energy (BLE) antenna just above the processor. If you tilt the micro:bit just right in the light, you can see the square-wave-like design embedded in the board in the top-left corner. This antenna allows the board to communicate with any other Bluetooth-enabled objects less than 100 meters away, according to the published specifications. The BLE, also called Bluetooth Smart protocol, enables a data rate over the air of 1 to 3 megabits per second, all while using less than 15 milliamps. Not only does this allow you to flash your board remotely with a laptop or smartphone, but it also lets you send sensor data from the board to another device without having to worry about draining your batteries. BLE is supposed to allow you to operate your device for weeks or even months using only a simple coin cell battery.

What's the Range of BLE?

Although the published specs for Bluetooth Low-Energy state an operational range of 100 meters, my editor and I had doubts as to what the *actual* range of these devices is, so I decided to conduct a few informal tests. For both tests, I used a "Find my phone" application that requires pairing the micro:bit with your phone. The application loads a script onto the board that asks you to press the left (A) button. When you do that, it sends a Bluetooth signal to the phone, and the phone hollers "Yoo-hoo! Here I am!" at you until you press an acknowledgment button. For the first test, I paired with my board and then walked through my house, seeing how far away I could get before the phone would no longer respond. The results were disappointing, to say the least: in a clear line-of-sight path, the phone lost the signal at 26 feet (about 8 meters) away. When I turned a corner, I immediately lost the signal *and* the phone disconnected from the micro:bit.

For the second test, I took the phone and the micro:bit to a local football field, where I wouldn't be dealing with local WiFi signals, walls, metal, and other possible interference, electromagnetic and otherwise. Again, I paired my phone with the micro:bit and walked away until the phone no longer responded to the device. The results? As disappointing as indoors. The farthest I was able to get before the phone no longer responded was again 26 feet. A few variables such as phone and micro:bit positioning seemed to affect results; holding the phone *one* way increased the range, whereas holding the micro:bit *another* way made communication completely impossible. Multiple attempts resulted in a maximum communication distance of about 8 meters.

The BLE specs probably apply to idealized conditions only—in a padded room, encased in a Faraday cage, etc. In addition, the size and shape of your antenna can make a real difference, and the antenna on the micro:bit is pretty small. I found some reports online of people getting ranges of over 200 meters, but I was unable to duplicate or confirm those results. In the real world it appears that you'll only be able to rely on Bluetooth connectivity when you're in the same room as your micro:bit. Keep that in mind as you design your future applications.

All right, that's the back of the board, where all of the behind-the-scenes action is. Let's take another look at the front (Figure 2.6).

FIGURE 2.6: Another look at the front of the micro:bit

The front of the micro:bit may be where all the magic happens when you're interacting with it, but there really isn't much there. There's a momentary push button on each side, A and B, each programmable by the user. Between them is a five-by-five matrix of low-power surface-mount LEDs, each of which is again programmable by you. These can be used to scroll text, display patterns, show arrows pointing in particular directions, and almost anything else you can think of doing with a grid of twenty-five tiny lights.

Along the bottom is the row of GPIO pins that we discussed earlier, though here you can see the labels for the most commonly used pins. As I mentioned in Chapter 1, the best way to access these pins is to purchase the edge-connector breakout board and simply slide your micro:bit into the slot, front-side up, as you see in Figure 2.7. This breakout board exposes a double row of pins more like the ones you're probably used to accessing on your Raspberry Pi board, and lets you use the header wires you probably already have in your toolkit. Be aware, however, that the number of pins is misleading; the pins are double-stacked, which means that each pair of adjacent pins leads to one single GPIO pin on the micro:bit. You do *not*, however, have to connect to both pins to interact with that GPIO pin. One or the other is sufficient.

So that's a tour around the little micro:bit board. It's a very basic device, designed to be easy to use and still be powerful enough to do interesting things. As a device for experimenters and hobbyists, it's a bit low-powered compared to the Raspberry Pi, but it also fills a completely different niche than the Pi and its ilk.

FIGURE 2.7: micro:bit inserted into edge connector breakout board

In the next chapter, I'll introduce you to the several different ways of programming the micro:bit and delve deeper into the programming environment I prefer—MicroPython.

What's with the ARM Processor Anyway?

If you've been keeping up with the embedded computer space, particularly in hobbyist devices like the Raspberry Pi and other single-board computers (SBCs), you're probably aware of how ARM processors seem to be taking over the market. The Raspberry Pi uses a 1.2 GHz quad-core ARM Cortex A53. The BeagleBoard uses a 1 GHz ARM Cortex A8, the micro:bit uses an ARM Cortex M0 and ARM Cortex M0+, and the list just goes on and on. Why are ARM processors so pervasive?

Today, 99 percent of smartphones and tablets have an ARM chip installed. The continuous drive to go smaller and lighter, and to pack ever more power into ever-smaller packages for cell phones has driven development of these chips at an extremely fast pace. The result—small, cheap processors— has had a huge impact on the SBC market.

The ARM (Acorn RISC Machine and then Advanced RISC Machine) architecture was introduced more than 30 years ago, in 1985. A British company called Acorn Computers brought out the BBC Micro, using its own 32-bit Reduced Instruction Set Computer (RISC) chip. The BBC Micro was hugely popular, going on to sell over a million and a half units, which helped to keep Acorn on the map. Although the successor to the Micro was kind of a flop, the concept of RISC machines was a winner, and Advanced RISC Machines was born as a separate research company in 1990 through a joint venture with Apple Computers and VLSI Technology. Advanced RISC Machines eventually became the ARM Holdings we know today. This graphic shows the Apple A5 chip, containing a dual-core ARM Cortex A-9.

ARM chips differ from Intel and AMD processors mainly in size and power usage. Mainstream CPUs are designed to efficiently "farm out" processing tasks to different devices on the motherboard, such as the GPU and the network interface card (NIC). ARM chips, on the other hand, have all of these things built into the chip. The ARM Cortex A-9, for instance, is a 1.2 GHz dual-core chip with onboard 3D graphics, 1080P video encoding and decoding, USB, PCIe and SATA interfaces, and various communication channels such as WiFi and GPS. It's not nearly as fast as a similar Intel or AMD processor, but it does what it does cheaply and inexpensively, power-wise.

Because it does these things so well, ARM chips can be found in an incredible variety of devices: game consoles, set-top boxes, personal media players, ebook readers, smart TVs, toys, coffeemakers, and a whole host of possibilities in your automobile, such as airbags, antilock brakes, computer management... the list just goes on. The continuing drive for smaller, more powerful devices means that manufacturers will continue to try to make chips more versatile, and ARM has proven itself to be a great platform for doing just that.

ARM's low power needs and continually growing architecture has also piqued the interest of the government; NASA and other government organizations like the Air Force are currently investigating using ARM processors aboard space vehicles (currently most satellites and probes use radiation-hardened Intel or PowerPC chips). NASA recently released a call for proposals for contractors to build a "chiplet"—a multicore ARM-based board that is easily extensible and could be used for future extraplanetary missions like the next Mars rover.

ARM Holdings was purchased by a Japanese group, SoftBank, in September 2016. This most likely will not affect consumer devices using ARM chips, but it may affect governmental contracts, because many devices used by the government are mandated by contract to be built using hardware and technology acquired from previously approved sources. We'll have to wait and see if this transfer affects ARM's takeover of space as well as the planet Earth.

3

Programming Using MicroPython

Now that you've had a very basic introduction to both the micro:bit's hardware and how to write a program to it, let's begin talking about how to actually write a program.

As of this writing, you can use at least six different coding environments to program your board: MicroPython, Microsoft PXT, Microsoft Block Editor, the Code Kingdoms JavaScript editor, Microsoft Touch Develop, and the mbed yotta integrated development environment (IDE). All of these languages have a JavaScript-based web environment you can use to create your programs, and for basic scripts that's probably all you'll need.

However, at some point you're probably going to want to dig deeper into the micro:bit's capabilities, and learning a more robust code is the best way to make that a bit easier.

I'm going to use this chapter to introduce you to Micro-Python, my language of choice when it comes to programming the micro:bit. You may know from my previous books that I'm already a big fan of Python as a coding language, so discovering that there was a native Python-esque way to interact with the board made it a no-brainer, in my opinion, as to what language to use and teach. If you've never used Python before, MicroPython is a great introduction to the language. It's a bit less feature-rich than full-fledged Python, but it's exactly the same, syntax-wise, which means that lessons learned in this environment will carry over well should you decide to learn regular Python as well.

We'll start with learning the web-based side of things, and then we'll explore how to program MicroPython for the micro:bit directly on your computer.

The MicroPython Web Editor

To start with programming via the web, point your browser to http://microbit.org/code/. Choose the Python editor and click the "Let's code" button. You'll be taken to a web-based editing page that looks like Figure 3.1.

As you can see, there's a (fully functional) example script already loaded for you. If you'd like to try it out, click the Download button. You'll be prompted to save the resulting hex file to your computer.

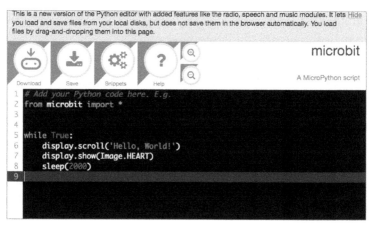

FIGURE 3.1: The MicroPython scripting page

Once it's downloaded, connect your micro:bit if you haven't already. It should show up as an external disc drive. Simply drag the downloaded hex file (`unearthly-script.hex` in this case) onto your micro:bit's drive icon. The board will flash, and then you should be greeted with "Hello, World!" scrolling across the display, followed by a heart symbol.

> **NOTE** If for some reason you plug your micro:bit into a USB slot on your computer and it lights up but doesn't show as an external drive, you may have a USB cable that's designed only for power, not data. They're not common, but they do exist. Before you return your board as defective, try another cable—it's likely to work.

A few icons are spaced across the top, so let's take a look at them. The first one, My Scripts, takes you to a page where you can see the scripts you've been working with since you visited the page. If you've signed in with a micro:bit account, this is where all your saved scripts will go. You don't need to create an account to use the editor, so creating one is up to you. The interface will save all your scripts until you close your browser, unless you've set your browser to save cookies; in that case, your scripts should remain saved indefinitely, which is nice. In any case, your scripts will save with names like *extraordinary script*, *unearthly script*, and so on. Unfortunately, there doesn't seem to be a way to rename your scripts unless you download them.

To the right of the Download icon is the Snippets button. This is less helpful than you'd think; I was hoping for some functional examples, such as accessing the buttons or onboard sensors for the board. Instead, the Snippets button shows you some common Python functions you can use in your code, like while, with, class, if/then, and so on. If you're new to the Python language, this can be useful, but if you're looking for more information about the micro:bit-specific functions, you'll want to skip ahead to the non-web-interface bit in the next section, where I discuss the mu programming environment. These snippets make the web interface act something like a regular IDE. Typing the first few letters of a Python keyword like import or if and then pressing the Tab key fills in the code for you. The interface also adds a helpful comment reminding you to fill it in (Figure 3.2).

FIGURE 3.2: Typing if and pressing Tab fills in the code for you.

Next to the Snippets button is the Help button, which takes you to (as you'd expect) an introductory web page—in another tab, luckily, so your current script is saved. This page walks you through the web-based interface, just like I'm doing now, and gives a few examples of some simple programs. In that sense, it provides more information than clicking the Snippets button.

Finally, there are two small icons on the right—the Zoom buttons. These let you enlarge or shrink the code, which is useful if you're displaying your screen with a projector for a group, for example.

It's important to remember that this *is* a fully functional programming interface, if a bit light in the documentation department. Once you're familiar with what you can do with your micro:bit and the commands and functions needed to program it, the web interface can be a handy way of throwing a quick script together wherever you may happen to be and quickly flashing your board. I think this interface would be most useful for teaching in a group setting, where students can follow along with a teacher's script and flash their boards without having to deal with installing a full IDE on their individual computers.

The mu Programming Environment

As I've already mentioned, the Python web interface is fine for some basic board programming, but there are some very good reasons for installing an IDE on your computer to do your programming. For one thing, you'll be able to save your work to your local machine, making it easier to work in stages. For another, code completion can be *extremely* helpful when you're just not sure what function you need or exactly what arguments are necessary when you're calling a function. Most good IDEs have a code-completion feature, where you can enter the first few letters of a word and the IDE will suggest common functions and already-created variables that will fit. In a situation like this one, where you're not sure what functions are available, code completion can come in *very* handy. The web interface has this functionality, but it's extremely limited in its vocabulary. For instance, in the web interface, typing `wh` and pressing the Tab key will fill in the block:

```
while True:
        # TODO: write code...
```

This is nice, but chances are that if you type `wh`, you already know you're planning to type `while` and the Tab completion is just a timesaver. However, if you type

```
display.show(Image.
```

in the web interface environment, nothing happens, while in the mu programming environment we're about to explore, the interface will clearly illustrate the possibilities that can follow `Image.` with a drop-down list.

As it turns out, there is a full IDE for MicroPython, and it's called *mu*. It's the basic development editor suggested by the micro:bit foundation for working with the board. It's an easy download and installation process, and it works for Windows, Mac, and Linux. Point your browser to http://codewith.mu and scroll to the bottom of the page, where you can choose your operating system flavor. Download and install the correct version; Windows uses an `.exe` file, Mac has a `.zip` file, and Linux uses a `.bin`.

If you're using a version of Windows earlier than Windows 10, after you run the EXE file, you'll need to install an additional driver in order to make a serial port available with which to communicate with your micro:bit. Click the link on the Download page, and follow the instructions included. According to the mu website, if you're using Windows 10 this step won't be necessary. However, when I tried to access the micro:bit using my own Windows 10 installation, I had to install that driver in order for things to work, so bear that in mind.

If you're using a Mac, download the `.zip` file to your local directory. When it's finished, right-click it and choose Open. This will result in a mu application (Figure 3.3) appearing in the same directory as the downloaded `.zip` file, which you can then copy to your Applications folder. Depending on your security settings, the first time you run it you may get an error because it's from an unsigned developer; if that's the case, just open the Security & Privacy settings in your System Preferences and let your computer know that it's all right to open the program.

mu

Application – 19.4 MB
Created Sunday, November 6, 2016 at 11:41 AM
Modified Sunday, November 6, 2016 at 11:41 AM
Last opened Sunday, March 5, 2017 at 5:11 PM
Version 0.0.0
Add Tags...

FIGURE 3.3: The mu application

If you're using a Linux box for your development, download the `.bin` file and save it to your local directory. Open a terminal, navigate to the directory you saved to, and type

```
chmod +x mu-0.9.13.linux.bin
```

(or whichever version you have downloaded) in order to make it executable. Then just type

```
./mu-0.9.13.linux.bin
```

to start the program.

According to the mu website, you may need to ensure that you're a member of the `dialout` group (in order to access the micro:bit over a serial connection.) To do that, in your terminal, type

```
getent group dialout
```

and you should see something like

```
dialout:x:20:<your username>
```

which tells you you're a member of the group. If your username is not listed, type

```
sudo adduser <your username> dialout
```

to add yourself. As an additional side note, on some Linux systems the serial port is owned by `serial`, not `dialout`. To check, type

```
ls -l /dev/tty*
```

and make sure that the owner of all devices listed is `dialout`, not `serial`.

Once you've installed mu, open it up, and you should see the window shown in Figure 3.4.

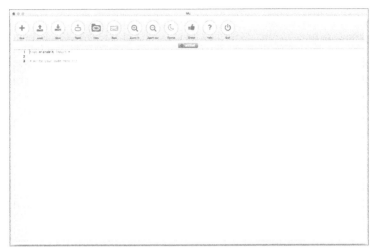

FIGURE 3.4: The mu editor

Now that you've got mu installed and open, let's take a look at the interface. It's designed to be very user-friendly and most of the icons are pretty self-explanatory. They also display a helpful dialog box if you hover your mouse over them. As you can see, some of them duplicate the web interface icons, since the web interface was designed to mirror the mu application.

On the left, the first group of three icons—New, Load, and Save—let you create, open, and save your script to your computer. Skipping the second group of three (we'll get to those in a moment), Zoom-in, Zoom-out, and Theme in the next trio let you enlarge or shrink the text size, and switch back and forth between a light or a dark environment. In the final group, Check lets you check your code for mistakes before you flash it to the micro:bit, Help opens the page https://codewith.mu/help/0.9.13/ with your default browser, and Quit quits the mu application completely.

Now let's return to that second group of icons: Flash, Files, and REPL. Flash is obvious—it loads the current working script onto the micro:bit board. Once you've got a working script in your window, just clicking the Flash button will load it onto your device. You should be aware of two things about this procedure, however. First, the code will be loaded onto your micro:bit whether or not you save it to your computer first. In other words, saving is not necessary to flash your board. Second, the code is uploaded without a prior check for errors. I say this because those among my readers who are familiar with using an Arduino may remember that the Upload button in the Arduino IDE first checks and compiles the code before uploading it to the Arduino. The mu interface, however, just uploads, errors or not. Because of this, it's a good idea to get in the habit of using the Check button before uploading a script. If (when) you *do* upload faulty code, the micro:bit tries to helpfully tell you what's wrong by scrolling messages across the board's LED display. However, you can imagine how difficult it is to read `Line 4 Name Error: Name 'shrubbery' is not defined` when it shows one hard-to-read character at a time, slowly scrolling past:

```
...L...
...i...
...n...
...e...
...4...
...N...
...a...
...m...
...e...
...E...
...r...
...r...
...o...
...r...
```

...and so on. *Please*, get in the habit of checking your code.

Files and REPL are a bit more involved than the Flash button. Files allows you to view files that are currently on the micro:bit. Clicking this icon brings up two additional panes at the bottom of your window (Figure 3.5).

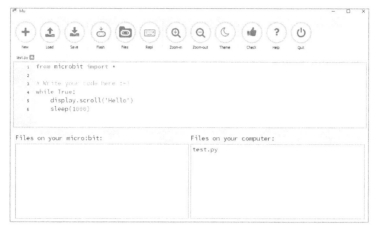

FIGURE 3.5: The Files interface

The left pane shows files available on your micro:bit, and the right pane shows files that are available in your local directory. If you've just opened mu for the first time and haven't saved any files, you won't see any files listed here. Try saving the open file as test.py and then close and reopen the Files window, and you should see test.py listed on the left.

REPL (Read-Eval-Print Loop) is perhaps the niftiest part of the mu IDE—it allows you to "live-program" your board, similar to an interactive Python session on your computer. To try it out, click the REPL button. (The Files interface and the REPL interface cannot be open at the same time; if you get an error message from the IDE, click the Files icon again to close the interface and then click REPL again.)

Your working window should split into two panes—an upper and a lower—where the lower pane is your interactive session. If you type help(), the display will scroll with a short introductory text, which gives you an idea of what you can do (Figure 3.6).

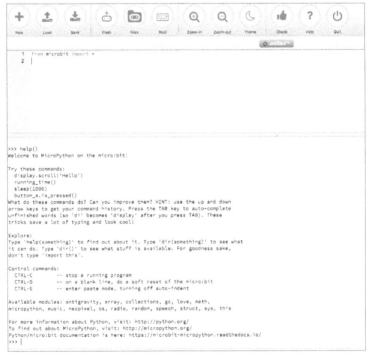

FIGURE 3.6: The REPL help() command

Try typing

```
display.scroll('Hello, World!')
```

at the prompt. Your micro:bit should immediately scroll "Hello, World!" across the LEDs on the front. Typing

```
display.show(Image.HAPPY)
```

should show a happy face. To wait two seconds, type

```
sleep(2000)
```

And finally, to clear the display of the happy face, just type

```
display.clear()
```

in the interactive window.

Feel free to play around with some different Python commands, but there's an important (in my opinion) aspect of this programming environment, aside from being able to immediately test commands: the `help()` interactive command. Similar to Linux's `man` pages, Python's `help()` function can be a lifesaver, especially in a situation like this one where you're not confident about how to use available functions. To illustrate, type **help(Image)** at the command prompt. The prompt immediately responds with a message about how to easily access each of the device's LEDs in a grid pattern with a command like

```
Image (
    '09090:'
    '99999:'
    '99999:'
    '09990:'
    '00900: ')
```

which will display a heart. Even more useful is the information it gives that you can control the brightness of each individual LED by using numbers from 0 (off) to 9 (brightest). To test this, try entering the following at the prompt:

```
>>> x = Image('02468:02468:02468:02468:02468:')
>>> display.show(x)
```

You should be rewarded with each row of your display showing LEDs of increased brightness as they go to the left. In fact, let's play with this newfound information. Close your REPL window by clicking the icon again, and then type the following code into your main window:

```
from microbit import *
a = Image('86420:86420:86420:86420:86420:')
b = Image('68642:68642:68642:68642:68642:')
c = Image('46864:46864:46864:46864:46864:')
d = Image('24686:24686:24686:24686:24686:')
e = Image('02468:02468:02468:02468:02468:')
while True:
    display.show(a)
    sleep(200)
    display.show(b)
    sleep(200)
    display.show(c)
    sleep(200)
    display.show(d)
    sleep(200)
    display.show(e)
    sleep(200)
    display.show(d)
    sleep(200)
    display.show(c)
    sleep(200)
    display.show(b)
    sleep(200)
```

Flash this to your micro:bit, and you should be rewarded with a line of dots moving back and forth across the display. (You can download this code from the GitHub repo here: https://github.com/wdonat/microbit-code/blob/master/chapter3/line.py.) You can also add this Python code directly to the online Python editor if you'd like to use it outside of the mu programming environment. Download the file from GitHub and then visit http://python.microbit.org/editor.html in your browser. Drag and drop the downloaded line.py file into the editor, press the Download button, save the line.hex file to your machine, and then flash it to your micro:bit board.

The code here is pretty simple. Variables a through d simply show a bright line at varying points on the display surrounded by progressively darker lines, and the while loop scrolls through each frame of the animation. There are two important lessons I should point out here, however. First of all, you may notice that you don't need to import time as you would in a normal Python script. The microbit module has a built-in method, sleep, that you call instead of time.sleep. It takes milliseconds as an argument, so sleep(200) tells the board to wait for 200 milliseconds before continuing. Second, there is a weird syntax requirement you may notice as you attempt to upload or check your script. After the final line of code, you'll need to press Enter/Return to create a newline, but then you can't have any whitespace in that newline. The compiler requires a newline/carriage return at the end of the last line of the script, but nothing *after* that. Since your last line is in the middle of a while loop, after your last carriage return you'll need to press Delete until your cursor is at the very beginning of the last line. It's an idiosyncrasy that you'll need to remember as you go along. You can flash code to your board with whitespace in the last line, but if you use mu's Check icon, it'll complain unless the whitespace is removed.

So there you have it: an introduction to writing Python code for (and directly to) your micro:bit. If you'd rather use a different coding language, I'll go through a few of the other possibilities in Appendix 2. Now that you're familiar with the programming environment, in the next chapter we'll discuss a few basic projects, like writing text to the display and accessing the onboard sensors.

Some Basic Projects

Now that you've had an introduction to programming the micro:bit in Python, it's time to start tackling some basic projects and getting used to interacting with the board and all of its inputs and outputs.

The LEDs

Probably the most visually interesting part of the micro:bit, obviously, is the 5×5 grid of bright red LEDs that take up the majority of the real estate on the front of the device. With these LEDs, you can print images, scroll text, and even play games. They are also important because we use them to display information (accelerometer data, instructions, and so forth) if the board is connected to an external device such as a smartphone.

You got a short introduction to using the LEDs in the previous chapter, but let's see what else we can do with them.

To start, make sure you have mu installed on your system. Plug in your micro:bit and open up mu.

Text

Since the built-in functions take care of turning individual LEDs on and off for you, probably the easiest thing to display on the board is text. There are several different ways to display text, but all of them use the display() method. In a blank script, try entering the following:

```
from microbit import *
while True:
    display.show('Hello, world!')
    sleep(200)
    display.show('Hello, world!', 200)
    sleep(200)
    display.scroll('Hello, world!')
    sleep(200)
    display.scroll('Hello, world!', 200)
    sleep(200)
```

Press the Check button to check for errors in your code (I know, it's a simple script, but it's good to get in the habit of checking—remember, mu doesn't check before uploading) and then flash the script to your board.

You'll notice that these lines of code—display.show() and display.scroll()—are just different ways of showing a line of text. The first three implementations of the show() function show the letters in the string, one after another, on the LED matrix. The show() function takes five parameters, only one of which (the actual text string) is required. The function's full parameters are

```
display.show(x, delay=400, wait=True, loop=False,
clear=False)
```

x is the string (or image, or character, or whatever) you want to display. delay (with a default value of 400) is the length of time in milliseconds between characters. wait (with a default value of True) determines whether or not the animation occurs in the background while other parts of the script continue (if `wait` is set to `True`, the rest of the script pauses while the animation takes place). `loop`'s default value of `False` means the text will only display once, and `clear`'s default value of `False` means the display will not empty after the last character displays.

The last few lines illustrate the `scroll()` method of displaying text, which—obviously—scrolls the letters across the matrix, marquee-style. Like `show()`, `scroll()` takes five parameters, only one of which is required. The function's full parameter list is

```
display.scroll(string, delay=150, wait=True,
loop=False, monospace=False)
```

`string` is obviously the string you want to display (`scroll()` does not work with images the way `show()` does). `delay` is the length of time it takes each letter to show (smaller numbers mean a faster scroll). Again, `wait` determines whether or not the animation occurs in the background. `loop`'s default value of `False` again means that the animation will not repeat. Setting `monospace` to `True` means that each letter will always take no more than five pixel-columns as it scrolls.

Images

If you want to show images on the micro:bit's LED matrix, the `display.show()` method is the one you'll be using. Recall that the first parameter of `show()`, `x`, can be a string, an image, a character, or even a list. This means that you can declare a list:

```
x = ['1', 'a', '2', 'spam', 'eggs', str(5 + 6)]
```

and then show each list member, one after another, with `display.show(x)`. A caveat here is that `show()` can only display either strings or what MicroPython recognizes as `Images`; `display.show(9)` will throw an error (`TypeError: not` an image), as will `display.show(Image.CLOVER)` (Attribute Error: type object 'MicroBitImage' has no attribute 'CLOVER').

That being said, however, quite a few built-in images are available for you to play with, and you can always create your own the way we did at the end of the last chapter. More images are sometimes added by MicroPython developers, but the ones currently available on the micro:bit run the gamut from clock displays 12:00 through 11:00 (Figure 4.1) to arrows pointing in all directions, to happy and angry faces, to check marks and houses and rabbits and snakes and even Pac-Man (Figure 4.2).

FIGURE 4.1: `Image.CLOCK2` (two o'clock)

FIGURE 4.2: `Image.PACMAN`

The best way to display a list of what's available is simply to start typing a line in mu and look at the code auto-fill suggestions. In your editor window, type

```
display.show(Image.
```

and take a look at the pulldown menu that becomes available (Figure 4.3) as soon as you type the period after `Image`.

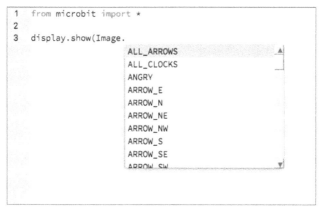

```
1    from microbit import *
2
3    display.show(Image.
                      ALL_ARROWS
                      ALL_CLOCKS
                      ANGRY
                      ARROW_E
                      ARROW_N
                      ARROW_NE
                      ARROW_NW
                      ARROW_S
                      ARROW_SE
                      ARROW_SW
```

FIGURE 4.3: Available premade images

If you want to create an image that doesn't appear in that list of ready-made pictures, creating one is not difficult. Probably the easiest is to use a matrix, the way we did at the end of Chapter 3, "Programming Using MicroPython." You're creating a `MicroBit.Image` object, which takes as a parameter a colon-separated list of pixel values, row by row. Those values can range from 0 (the faintest) to 9 (the brightest). In other words, you can create a square with an embedded X with the following commands in your REPL window:

```
>>> a = Image('99999:99099:90909:99099:99999:')
>>> display.show(a)
```

Figure 4.4 shows the result. In fact, if you're interested, you can see exactly how each of the ready-made pictures are constructed by simply typing the name of the image in your command prompt and looking at the response. For example, typing **Image.PACMAN** returns

```
Image('09999:99090:99900:99990:09999:')
```

It's not terribly high-resolution, but by fiddling with the brightness of each individual pixel, you'd be surprised at just how much control you can have when it comes to creating images and animations.

FIGURE 4.4: Square with embedded X

For instance, creating an image with a bright pixel in the center, surrounded by gradually dimming pixels, and then "moving" that bright pixel around, can create a soothing effect. To see what I mean, download https://github.com/ wdonat/microbit-code/blob/master/chapter4/animation.py, paste it into your mu window, and flash it onto your board. You can see a single frame of it in Figure 4.5.

You'll see by looking at the code that you can create a list of images, and then use show() to animate that list with a specific delay. Setting loop equal to True ensures that it goes on forever.

FIGURE 4.5: A single frame of pixel animation

You can also access the values of individual pixels using `display.get_pixel(x, y)`. This function takes as a parameter the pixel location (column = x, row = y) and returns an integer from 0 to 9, 0 being `off` and 9 being the brightest.

That was a good introduction to creating text and images on your micro:bit. Now let's move on to using the buttons.

The Buttons

The buttons labeled A and B on the front of the micro:bit board are momentary pushbuttons that can be used as inputs for various reasons: gameplay, responding to

Getting Started with the micro:bit

prompts, or various other possibilities. They're accessed with MicroPython objects `button_a` and `button_b`, appropriately enough, and those objects have a few functions associated with them: `get_presses()`, `is_pressed()`, and `was_pressed()`.

To get a feel for how they work, start a new mu script and enter the following:

```
from microbit import *
while True:
    if button_a.is_pressed():
        display.show(Image.HAPPY)
        sleep(500)
        display.clear()
    if button_b.is_pressed():
        display.show(Image.ANGRY)
        sleep(500)
        display.clear()
```

Save this and flash it to your micro:bit. As you've probably guessed, it simply responds to your button press with either a happy or an angry face, depending on which button you've pressed.

You can also keep track of things such as button presses using the `get_presses()` method. Again, in your mu window, try the following:

```
from microbit import *
while True:
    display.scroll("Press the A button a few times.")
    sleep(5000)
    display.scroll(str(button_a.get_presses()))
```

This is a remarkably simple script—while the board sleeps for five seconds, it counts how many times you press the A button, and then displays the count. Accessing the buttons on the micro:bit is pretty easy—there are only three functions to call, after all. Now let's take a look at the sensors on board.

The Accelerometer

The micro:bit's onboard accelerometer is one of the more nifty things about the board, because you can do a lot with it. It measures in three axes, which means it can sense movement about the X-axis (horizontal left and right), the Y-axis (horizontal forward and backward), and the Z-axis (up and down). You can read individual X-, Y-, and Z-values, or you can recognize gestures. The micro:bit recognizes eleven different gestures, all as strings: up, down, left, right, face up, face down, freefall, 3g, 6g, 8g, and shake. Most of these are self-explanatory, but 3g, 6g, and 8g are not. Basically, these gestures are registered when the board experiences those levels of G-force.

The G-force gestures may be difficult to experiment with (unless you feel like throwing your micro:bit against the wall), but we can use the seven built-in accelerometer methods to look at the other values that the device can send. These methods are get_x(), get_y(), get_z(), get_values(), current_gesture(), is_gesture(), was_gesture(), and get_gestures().

Let's look at a few of those. With mu open and your board connected, click the REPL icon and connect to your board. At the prompt, type

```
>>> accelerometer.get_values()
```

You should be rewarded with a Python tuple (a comma-separated list of values in parentheses) of three values—an X-component, a Y-component, and a Z-component, similar to this:

```
(272, 224, -960)
```

Moving the board around and repeating the command should show you the differing values as you move it. These values are in milli-g's—thousandths of a G.

At another prompt, type

```
>>> accelerometer.current_gesture() == "face up"
```

This *should* return either True or False, depending on your board's current orientation. However, you may get False, no matter how you maneuver the board. This may be a bug in the interactive Python interface; if this happens to you, don't despair. Just try the script in the next paragraph to make sure your accelerometer is working the way it should.

This command, by the way, is an example of a Boolean conditional in Python. You could use this command in an if statement in a Python script, and it will return either True or False. You could then instruct your micro:bit to act accordingly. For instance, close your REPL window by clicking the icon, and enter the following code into your IDE window:

```
from microbit import *
while True:
    gesture = accelerometer.current_gesture()
    if gesture == "face up":
        display.show(Image.HAPPY)
    else:
        display.show(Image.ANGRY)
```

Flash this short program to your board. This script comes straight from the micro:bit MicroPython documentation on GitHub, and it simply tells your micro:bit to smile if it's face up and frown if it's not (Figure 4.6).

FIGURE 4.6: micro:bit unhappy at not being face up

Another useful function is `was_gesture()`, which lets you know if a certain gesture has been performed. To try this out, type the following into your mu window and flash it to your board:

```
from microbit import *
while True:
    display.show(Image.DIAMOND)
    if accelerometer.was_gesture("shake"):
        display.clear()
        sleep(1000)
        display.show(Image.YES)
        sleep(1000)
```

This script will make the micro:bit display a diamond shape until you shake it; at that point the screen will clear for a second and then display a check mark. Fair warning: It may be easier to test this script if you use the external battery pack or have your micro:bit hooked up to the mi:power board, because it's rather difficult to shake the device while it's connected to your computer via USB cable.

If you'd like to experiment further with the accelerometer and its readings, you can also use the REPL window. For instance, entering

```
>>> display.show(str(acccelerometer.get_x()))
```

will show the current X-value on the LED display. (Note that you have to cast the `accelerometer.get_x()` reading to a `str`, since the `display.show()` function only takes strings, and the `get_x()` function returns an integer.) You can also play with the `3g` and other `-g` functions, but as I said, be careful about breaking your board!

The Compass

The last sensor on the micro:bit board we want to play around with is the compass/magnetometer. It works by detecting magnetic fields and determining their direction. Before you use it, you'll need to calibrate it using a built-in function called—appropriately—`calibrate()`. You can call this function interactively at the REPL prompt, and you can also call it in a script; if it's called in a normal script, the script will pause while the calibration takes place. The calibration itself consists of a little game in which you have to draw a circle by tilting the board, using a single LED on the display (Figure 4.7).

FIGURE 4.7: Calibrating the compass

Once the circle is drawn and you've calibrated the compass, you're greeted by a smiley face (Image.HAPPY). Try it out:

```
>>> compass.calibrate()
```

Once you play the game, you're done. The device should stay accurate as long as it's powered on, but a power cycle may destroy the accuracy. That's why it's a good idea to add the compass.calibrate() line to any script that requires the compass—doing so will ensure everything is accurate before progressing further in the program.

The board has a total of eight compass-related functions: calibrate(), clear_calibration(), get_field_strength(), get_x(), get_y(), get_z(), heading(), and is_calibrated() (Figure 4.8).

```
1    from microbit import *
2
3    compass.
              calibrate
              clear_calibration
              get_field_strength
              get_x
              get_y
              get_z
              heading
              is_calibrated
```

Of these, `compass.heading()` is one of the easiest to understand. Point the top of the compass (where the USB plug is) toward North and type **compass.heading()** in your REPL prompt, and you should be rewarded with a heading in degrees (from 1 through 359). Since you're pointed toward North, you should get a reading that hovers around 0 or 360 (depending on how you're pointing it exactly). If you get something completely different, recalibrate and try again. If you *still* get a value that's totally off base, check to see if you're holding your micro:bit over a big magnet (Figure 4.9). Since the compass is nothing more than a Hall effect magnetometer, the presence of a strong magnetic field will adversely affect your directional readings. It is also affected by the presence of ferromagnetic materials around the device, so moving to a different location may solve the problem. Luckily, magnetometers are not affected much by electromagnetic (EM) waves, so you shouldn't have to worry about your television or microwave screwing up your results.

You may also need to play with your board's orientation. One of my boards was accurate when the LED display was facing up. The other, however, was only accurate when the display was facing down. Your results may vary.

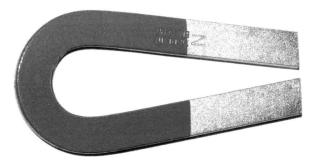

FIGURE 4.9: This will throw your compass readings *way* off.

get_x(), get_y(), and get_z() give the magnitude of the magnetic field in each axis, respectively, with the positive or negative depending on the direction of the field, and get_field_strength() returns the total magnitude around the device. All of these readings are returned in nanoteslas. Earth's magnetic field at ground level varies from about 25 to 65 microteslas, so the return of compass.get_field_strength() should be between 25,000 and 65,000 (microteslas converted to nanoteslas), assuming you're not holding your board over a horseshoe magnet.

The Local Persistent Filesystem

The last not-so-basic feature of the micro:bit I'd like to introduce you to is its filesystem. Yes, you can actually store files on the board—both text files and even binary files such as images. There are two caveats to this, however:

- ☑ There is only approximately 30 KB of memory on the board. To put that into perspective, the image in Figure 4.10 is 400×300 pixels, and is 31 KB in size. Now, it's true that you can get a lot of text into 30 KB, but my point is that you shouldn't plan on being able to store *War and Peace* on your micro:bit.

FIGURE 4.10: A 31 KB JPEG image

▰ Files on the board will remain only as long as you don't flash the board with a new program. Flashing the board necessarily wipes all onboard memory, so any files you have stored will be wiped out. You can safely turn it off and on again and the files will be safe, but once you flash it, they're gone forever.

That being said, you can imagine that the ability to store files on the board can come in handy. You can write to a text file for reading during a game, for example, or store the current configuration of LEDs as a text array or even an image file that you've converted. There is no directory structure; files are simply stored as a list. In order to work with the files, you perform an `import os` in your script and use the included functions (`open()`, `listdir()`, and so on) to operate on them.

As an example, open your REPL window in the mu editor and try the following:

```
>>> import os
>>> os.listdir()
[]
```

You should get an empty pair of brackets (`[]`) as evidence that nothing is stored on the device. To continue with experimentation:

```
>>> with open('test.txt', 'w') as f:
...     f.write("Hello, world")
>>> os.listdir()
['test.txt']
```

If you're still a Python newbie, I should probably mention here that after you type the first line starting with `with`, the mu interpreter will continue to indent the following lines until you tell it to stop. To do that, hit the Backspace or Delete key, and then Enter.

And finally, to read what you've written:

```
>>> with open('test.txt', 'r') as f:
...     print (f.read())
Hello, world
```

(Again, use the Backspace key to break out of the loop.) If you're familiar with Python, you should recognize the `open()`, `write()`, and `read()` functions, as well as the `with()` method. When you call the `open()` function, the second parameter determines whether you'll be reading from (`'r'`) or writing to (`'w'`) the file, as well as whether it will be stored as text or binary. The default is text, or `'t'`; if you want to store binary as bytes, the syntax uses `'b'`. If you want to read bytes from a binary file, for example, you would use `'rb'` as your second parameter to the `open()` function.

Obviously you don't always have to use the `os.listdir()` function to see what's on the device. You may remember that the Files icon at the top of the mu editor lists the files as well. However, be aware that this can still be a little buggy; nine times out of ten when I clicked the Files icon on my computer (a Mac running the Sierra OS) the interface would freeze and

I would have to force-quit the program. `os.listdir()`, how-ever, has never crashed for me.

Putting It All Together

Now that you've had a chance to play with all the various parts of the board, it's time to experiment with putting things together in some projects. The easiest way to play with all the buttons and doodads on the board seems to be, appropriately enough, with games. Let's start with craps.

You're probably familiar with the game of dice from either playing it yourself or watching it in movies. At its core it's incredibly simple: roll the dice and bet on the outcome. All players bet on the same roll—the only thing that matters is the total of the dice, and each player rolls until he or she rolls a seven, which is called "sevening-out."

Now, we don't have to go over the betting; I'll leave that to you. What we *can* do, however, is simulate the dice rolls, and—just to make it interesting—enable a way for us to cheat, using the board's two buttons.

The basic structure of the script will be as follows:

- Shake the micro:bit.

- This will result in two numbers being chosen at ran-dom, which we'll display after the roll.

- Cheat #1: Holding down the A button while you shake will result in a roll of 7, enabling you to pass the dice if you don't want to roll anymore.

- Cheat #2: Holding down the B button while you shake will result in a roll of 8 (or whatever you choose), ensuring that you can bet on a known outcome.

I would like to emphasize here that this is for instructional purposes only, and is probably not a good way to study statistics and probability. It is all for fun, after all.

So let's write this simple program. Open up mu on your computer and enter the following:

```
from microbit import *
import random
random.seed()
```

These lines import the microbit libraries and the random number generator. The `random.seed()` line is necessary, because it "seeds" the random number generator. This ensures that the numbers chosen will indeed be random (or at least as random as is possible without invoking quantum computations). Let's continue:

```
def roll_dice():
    a = random.randint(1, 6)
    b = random.randint(1, 6)
    numbers = [str(a), str(b)]
    return numbers
def roll_seven():
    a = 4
    b = 3
    numbers = [str(a), str(b)]
    return numbers
def roll_eight():
    a = 5
    b = 3
    numbers = [str(a), str(b)]
    return numbers
def show_roll(dice):
    for i in range(0, len(dice)):
        display.clear()
        display.show(dice[i])
        sleep(1000)
    display.show(Image.YES)
```

These are just three functions that simulate dice rolls and one that shows what was rolled. Which dice roll gets called will depend on whether a button is being held down during the shake. Each function selects a value for each die, and then returns a list of those two numbers, converted to strings for easy displaying. Finally, the show_roll() function takes a list, dice, as its parameter. It simply shows each member of the list in order, twice, to make sure you can see the roll.

Now comes the main portion of the program:

```
display.show(Image.YES)
while True:
    if not accelerometer.was_gesture("shake"):
        continue
    if button_a.is_pressed():
        roll = roll_seven()
        show_roll(roll)
        continue
    if button_b.is_pressed():
        roll = roll_eight()
        show_roll(roll)
        continue
    roll = roll_dice()
    show_roll(roll)
```

As you can see, this is a very simple program. (You can download it at https://github.com/wdonat/microbit-code/blob/master/chapter4/craps.py.) The default image shown on the LED screen is the YES symbol (a check mark). As soon as a shake is detected, the device checks to see if a button was held down as well, and then calls the appropriate function, which clears the screen and displays the die values. Then it waits for a second and returns to the default image.

Lastly, let's just write a short script that walks through the available functions on the device, just to make sure we know how to access them and that our device is working properly. Again, this should be a simple program:

```
from microbit import *
compass.calibrate()
```

Now you're ready for the function that will go through all of the functionality of the board:

```python
def iterate_thru():
    display.show(Image.ALL_ARROWS)
    display.clear()
    display.scroll("Press A")
    while not button_a.is_pressed(): # see text
        continue
    display.show(Image.ARROW_W) # see text
    sleep(1000)
    display.scroll("Press B")
    while not button_b.is_pressed(): # see text
        continue
    display.show(Image.ARROW_E) # see text
    sleep(1000)
    display.clear()
    x = str(accelerometer.get_x())
    y = str(accelerometer.get_y())
    z = str(accelerometer.get_z())
    a = x + " " + y + " " + z # see text
    display.show(a)
    sleep(1000)
    display.clear()
    display.scroll(str(compass.heading()))
    sleep(1000)
    if accelerometer.is_gesture("face up"):
        display.show(Image.HAPPY)
    return
```

And finally, the main loop of the program itself:

```python
while True:
    display.clear()
    if button_a.is_pressed():
        iterate_thru()
```

You can download this program at https://github.com/wdonat/microbit-code/blob/master/chapter4/functions.py.

While simple, there are a few things worth mentioning in this program, delineated by the see text comments:

- ☑ The while not loops make the board wait until button A (or B) is pressed by entering a loop; as long as the button is not pressed, the loop continues indefinitely. The button press breaks it out of the loop and allows the program to continue.

- ▧ `Image.ARROW_W` and `Image.ARROW_E` are probably self-explanatory, but in case you weren't sure: they point to the left and the right, respectively, as if you were looking at a map. Remember, the board thinks that the USB port is on the North side, so it is prudent to continue that framework with an East and a West arrow. Since there is no `ARROW_RIGHT` or `ARROW_LEFT`, these are what you would use instead.

- ▧ Finally, the assignment of the x, y, and z strings to the a string is due to the unique requirements of `display.scroll()` and `display.show()`. Both require strings, or images, or lists of images. A list of strings won't work. For that reason, we've combined the x, y, and z strings into one string, which can then be scrolled through.

That concludes our basic introduction to programming the integral parts of the micro:bit with Python. In the next chapter, we'll dive into interfacing with the board using C++ and the native tools that are available for that.

5

The mbed Operating System

Now that you're familiar with MicroPython and using the mu interface to interact with your micro:bit board, I'd like to sweep the rug out from under you and introduce the ARM mbed interface.

I know, it's harsh, and I can hear some of you screaming from here: "Wait! I *like* Python! Why do I have to learn something different?" The reason lies in the fact that mbed is the actual *operating system* that is running on the micro:bit's ARM chip, and in order to do really cool and effective things with that chip, you'll need to interact with the operating system in the most efficient way possible: with mbed and its language of choice, C++. Don't get me wrong—you can still use Python and mu to do many, many things with your board without an mbed account; it's just that learning to use mbed will enable you to do many, many more.

mbed OS 5 (the current version) is an operating system that is specifically designed to run on ARM microcontrollers and enable them to interact with the Internet of Things (IoT). If you'll remember, interfacing with the IoT is one of the stated goals of the micro:bit and its foundation. mbed was designed by ARM and its partners, and as of this writing it works on well over a hundred different ARM-based boards. Interacting with it is simple and will make you feel right at home. After applying for a developer account, you can use the web-based IDE to write your programs, compile them, and then flash them to your micro:bit board. Yes, you'll need to use C++, but it may not be as bad as you fear. I'll walk you through the worst of it, and remember that if you decide to continue programming, C++ is an excellent language to know. You can also download the mbed tools to your computer if you prefer to work that way. But installing and using those tools is slightly beyond the scope of this book, so we're going to stick with the online version of the compiler.

Getting an Account

Let's start with getting an mbed developer account, which you'll need to have in order to use the IDE. Start by pointing your browser to https://developer.mbed.org/ and clicking Log In/Signup at the top of the screen. Now click the big Sign Up button in the middle of the screen and follow the prompts to create a developer account (Figure 5.1). Don't worry—everything is free.

Once you have an account, you'll be sent to your dashboard, which is where you'll receive notifications of code updates and blog entries by the mbed team, as well as information about any teams you may be involved in, your repositories, and outgoing pull requests. With your new account, your dashboard should be empty, with the possible exception of some notifications from the mbed team.

To start writing code, click the Compiler button next to your username in the top menu bar. The mbed compiler workspace will open in a new window (Figure 5.2).

FIGURE 5.1: Signing up for an mbed developer account

FIGURE 5.2: The mbed compiler workspace

If you've worked with compilers and IDEs before, this setup may seem familiar, with the code/source tree on the left and the main coding window in the middle.

The first thing you'll need to do is let the compiler know what board you're going to be developing for by selecting a device. You'll notice that at the top right of the window it says No Device Selected. That's actually a button—click on it and a window will pop up, enabling you to select an mbed-enabled platform. Click the Add Platform button and select the BBC micro:bit on the next page. To make it easier to find, you can filter your search on the left of the page and check the box next to BBC Make It Digital Campaign under Platform Vendor. Click on the picture of the board, and you'll be taken to a description page for the micro:bit. Scroll a little way down the page and click the orange Add To Your mbed Compiler button on the right.

Switch back to your compiler page for a moment, and click the No Device Selected button again. You'll now see that the micro:bit is available as a device. Choose Select Platform, and back on the compiler page you'll see that the micro:bit is now showing as your selected device.

Switch back to the description page for the micro:bit for a moment. Beneath the pinout diagram of the board are several code examples for the board that you can immediately import into your compiler and play with, ranging from the usual `microbit-hello-world`, to more complicated code for advanced users.

Let's start with the `microbit-hello-world` program. Click the Import Program button next to the name of the code. A new window to your compiler will open, and a pop-up in the center will ask you for details about importing the program (Figure 5.3).

Leave everything set to the defaults and click Import. You'll see that `microbit-hello-world` is now part of the code tree on the left, and the main window is populated with a `microbit` folder and a `main.cpp` file. Double-click the `main.cpp` file, and the code window should now show you the file.

FIGURE 5.3: **Importing** `microbit-hello-world`

It's really a simple program—only seventeen lines if you don't count the extensive comments at the head of the program. All the important, behind-the-scenes code is included in the `microbit` folder. In your `main.cpp` file, the line `#include "MicroBit.h"` pulls in all the relevant code. (See the "Getting More Help" sidebar for help with learning C++ code.)

So let's flash this code to your board. Make sure your board is connected to your computer, double-check that the BBC micro:bit is still displayed as your compilation target at the top right of your window, and click Compile in the menu bar of your workspace. A status window will pop up, letting you know what's going on, and then either a hex file will download automatically to your default downloads folder, or you'll get a download prompt. You'll also notice that the status window at the bottom of the screen reads Success! in the Compile Output tab. If you have errors or warnings when compiling code, this window is where they'll display so you can work your way through them as you debug. Once you've downloaded the hex file to your computer, drag and drop it onto your board like you did before, and your micro:bit should scroll "HELLO WORLD! :)" across the LED display once.

If your only exposure to programming thus far has been Python, I understand if you're a little apprehensive about poking around the micro:bit with C++. It can seem a little daunting. However, if you've been playing around with Python, you should already have a grasp of how programming languages work and the logic involved in programming, which is—in my opinion—half of the battle. Getting the logic down for a program is the hard part. After that it's just a matter of getting it into the correct syntax for your situation. And though this book is not the place to familiarize yourself with C++, I'll do what I can to make it easy for you.

The first suggestion I have is to pick up a good Learn C++ book—there are literally thousands of them out there. I'm partial to *C++ All-in-One For Dummies*, by John Paul Mueller and Jeff Cogswell (For Dummies, 2014), but choose one that works for you.

Let's talk about ways to immediately learn more about classes and functions you'll be using. These are similar to Python's `import` statements. Various functions and external pieces of code can be included in another file, often ending in `.h`, and the `#include` statement makes that code available to your `main.cpp` program. We won't be creating these `.h` files (called header files) here, but you will be `#including` them from the mbed libraries.

If you've imported the mbed_blinky example code into your compilation workspace, click the plus sign next to the `mbed` folder to view its contents. You'll be greeted with three new folders of documents: `Classes`, `Structs`, and `Groups`. Expanding any of these will give you more information than you're probably ready for about the programming involved behind the scenes in the micro:bit. However, this is also a useful place for documentation if you don't know what programming structure to use or how to make use of it. In that sense, it's similar to Linux's `man` pages, or the `help()` functionality in Python and the REPL interface in mu.

In `Classes`, for example, click the `DigitalOut` document, since the `DigitalOut` class is used in the code in `main.cpp`. You'll see the class reference, how to use it (`#include <DigitalOut.h>`), and its functions and parameters (see the following graphic).

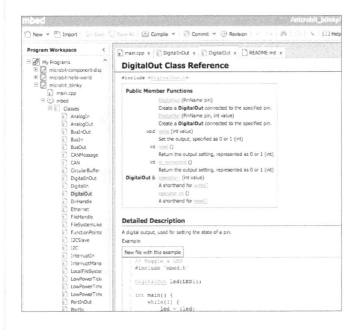

Following the introduction, there's a ton of information, including examples, on how to call the class, and this is perhaps the best way to get acquainted with the C++ code you'll be using. For example, scroll down and take a look at the Constructor and Destructor Documentation. If you look at your `main.cpp` code again, you'll see that line 8 reads

```
DigitalOut col0(P0_4, 0);
```

which corresponds to the `DigitalOut` constructor that takes a `PinName` and an `int` as parameters. This tells you that line 8 is creating a `DigitalOut` class with an initial value of 0 connected to the pin with `PinName` P0_4.

Obviously, going into all of the functions and formats in the micro:bit source code is beyond the scope of this book, but just know that there is help and documentation available. And in all of the projects in this book, I'll do my best to make sure you know what exactly you're programming and how and why it does what it does.

Welcome to C++!

When you've got that working—and you're tired of watching it—let's try another, slightly more complex piece of code: the micro:bit "blinky." If you've ever used an Arduino, you're probably familiar with the Blink sketch that's often used as a Hello, World introduction to the programming environment and to ensure that the board is connected and working properly. The mbed IDE has a similar program, called *blinky*, that can be downloaded and compiled for any of the 117 different boards that are running the mbed OS.

Unfortunately, the default program won't work if you compile it for the micro:bit, and that's because the micro:bit doesn't have a single LED that you can flash on and off. Rather, all of the LEDs on the board are part of the LED matrix, so you have to address an individual LED on the board in a slightly different way. Back on the micro:bit description page, locate the microbit_blinky program description. Click on it, and then click Import Into Compiler on the next repository page.

When you've imported the program into your compiler, click on `main.cpp` and take a look at the program. Again, it's a simple program, but the comments illustrate how you have to interface with the LED matrix in order to address just one LED. It's a bit more complicated than simply sending a digital HIGH or LOW to a single LED—at least at first. However, once you've set column 0 to be permanently at ground, the code does break down to exactly that—sending a digital signal to the chosen LED (in this case, `myled`, defined as `P0_13` in line 10).

This is the procedure you'll follow for your future micro:bit programs. We'll use the micro:bit samples program as a template, and just edit the `main.cpp` file as we need to.

yotta

Now that you're familiar with the online compiler for mbed, I'd also like to introduce you to *yotta*, mbed's command-line, Python-based tool that allows you to develop both source code and reusable modules on your local system. You can use it to build tools for your local Unix-based machine, as well as any of mbed's ARM-based boards; all you have to do is specify the build target when you compile your code. We'll be using the web-based tool for the remainder of this book, but I feel I'd be remiss if I didn't at least quickly walk you through installing and using yotta, since it's really a pretty cool tool. It's also extremely community-based; mbed states that the most important part of yotta is the ability to publish your module to make it available for other users. You can't publish executables, since it isn't really designed for that, but the source code you publish can be compiled for whatever target another user wishes.

As you read through this section, please keep in mind two things. First, yotta is very much a tool in beta. Things are likely to break, and it can be difficult to set up a working build environment. Second, with that being said, it is not necessary for you to get this working in order to continue with the book. All future projects can be done using the web-based interface, so if you can't get yotta working or don't want to spend the time on it, that's perfectly all right. You can do all sorts of cool things with your board without ever touching yotta.

Installing yotta

Installation differs, obviously, depending on your operating system.

Windows

It's important to remember that yotta requires Python to run, and Windows (still!) doesn't come with Python preinstalled. If you've skipped all the previous parts of this book and still haven't installed Python, head over to http://python.org/downloads and download and install the correct version for your system. You need at least version 2.7.9 for yotta to work. The all-in-one installer does install Python for you, however, so you can go that route if you prefer.

To install yotta using the all-in-one tool (definitely the easiest option) point your browser to https://github.com/ARMmbed/yotta_windows_installer/releases and download the yotta_install_v0.2.3.exe file. Double-click the downloaded file and follow the prompts. It'll take a while to install, but when it's done you'll have a Run Yotta icon on your desktop. Double-click it to bring up the command-line interface (Figure 5.4).

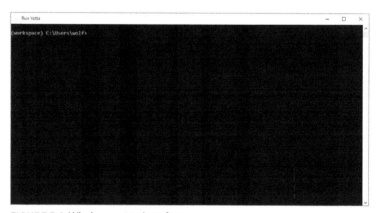

FIGURE 5.4: Windows yotta interface

macOS

The Mac version of yotta is currently at version 0.0.4, and it can be found at https://github.com/ARMmbed/yotta_osx_installer/releases/tag/v0.0.4. Download the `.dmg` file and drag the yotta app into your Applications folder from within the expanded `.dmg` folder. Once it's finished copying and verifying (warning: this can be a *looooong* process so go get yourself a cup of coffee or tea while it does this), double-click the yotta icon in your `Applications` folder. You'll be greeted by a terminal-like window (Figure 5.5).

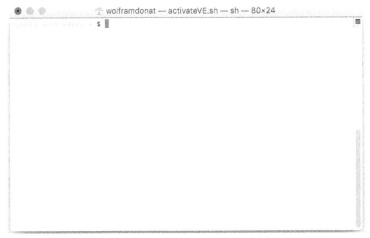

FIGURE 5.5: **The macOS yotta window**

Linux

A Linux installation can be done with your package manager and Python's `pip` tool. First make sure you have the correct dependencies on your system. On Debian and Ubuntu, update your package list and then use `sudo apt-get install` to install.

- ☑ `python-setuptools`

- ☑ `cmake`

- build-essential

- ninja-build

- python-dev

- libffi-dev

- libssl-dev

- srecord

Then run `sudo easy_install pip` to install the `pip` tool.
Fedora is a similar process. Install the dependencies:

```
sudo yum install python-pip cmake ninja-build python-
devel libffi-devel openssl-devel srecord clang
sudo yum groupinstall "Development Tools" "Development
Libraries"
```

The version of `pip` you get here is most likely out of date,
so update `pip` with

```
sudo yum remove python-pip
curl -o get-pip.py https://bootstrap.pypa.io/get-pip.py
sudo python get-pip.py
```

Now that you're ready, use `pip` to install yotta:

```
pip install yotta
```

After the Installation: Customizing yotta

Once yotta is installed, you'll need to set up a cross-compilation environment so you can develop for your board.
What this means, basically, is that after you write your code,
you tell the compiler that you don't want to compile the
code to run on the current computer (ordinarily the default
behavior). Rather, you want the compiler to compile the
code with a different *target* in mind—the micro:bit board.
Again, this process differs slightly depending on your host
system's OS.

Windows

The first thing you'll need to install on your Windows machine is the gcc compiler. You can download it from

https://launchpad.net/gcc-arm-embedded/4.9/4.9-2015-q2-update/+download/gcc-arm-none-eabi-4_9-2015q2-20150609-win32.exe

After it's installed, you'll need to add the `bin/` subdirectory from where you installed it to your environment path. This differs slightly depending on your version of Windows, but in general you'll right-click Computer and select Properties. Click Advanced System Settings, choose the Advanced tab, and click the Environment Variables button. Then select Path from the System Variables window (Figure 5.6) and add the `/bin` subdirectory to the end of the line, using a semicolon first with no spaces. Save everything and exit.

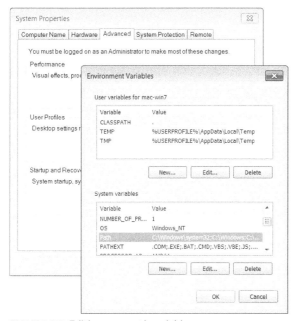

FIGURE 5.6: Editing your path variable

The last thing you'll need to do before compiling is to set your target. Close and reopen your yotta build command windows (to ensure the cross-compiler is available) and type the following:

```
yotta target bbc-microbit-classic-gcc
```

Because the micro:bit library isn't hosted by mbed, you'll need to download it from GitHub. To download the library, enter the following:

```
yotta install lancaster-university/microbit
```

The first time you do this, you may need to log in to your mbed developer account, so make sure you've verified your email address beforehand, as it may need to be verified before you can download the library.

You need to do this only once; all future commands will use that information when compiling.

macOS

As with Windows, to cross-compile, you'll need the arm-none-eabi-gcc cross-compiler. The easiest way to install this is using homebrew (see the sidebar "Homebrew on the Mac"). First add (tap) the ARMmbed homebrew package repository with

```
brew tap ARMmbed/homebrew-formulae
```

Then install the compiler with

```
brew install arm-none-eabi-gcc
```

Finally, set the target for the compiler to be your micro:bit board. Close and reopen any yotta windows you have, and then enter

```
yotta target bbc-microbit-classic-gcc
```

Follow that with

```
yotta install lancaster-university/microbit
```

and you should be ready to go. The first time you do this, you may need to log in to your mbed developer account, so make sure you've verified your email address beforehand, since it might need to be verified before you can download the library. You'll need to do this only once—yotta will remember your info when you download libraries in the future.

As a recent convert to the homebrew package management system, I thought a quick note about it might be worth something.

I'm mainly a Linux guy; I do all of my development work, both hobby and professional, in various flavors of Linux. However, my personal computer is a Mac, and I've always wished that I could use Linux's aptitude package manager on my Mac. It can't be beat—need a software package? Just type `sudo apt-get install <package>` and aptitude will find the package, determine its dependencies, and install everything for you. I wasn't aware of anything similar for the Mac.

Until I found homebrew. I'd heard of it but had never tried it. Now, I won't go back. It's the package manager that macOS needed, and it's very easy to install and use. First, point your browser to http://brew.sh, where you can read all about it. But installing it is a simple command-line entry:

```
/usr/bin/ruby -e "$(curl -fsSL
https://raw.githubusercontent.com/Homebrew/
install/master/install)"
```

Once it's installed, you can add repositories with `brew tap <repository name>`, and then install software packages from them with `brew install <software package>`. It's almost as easy as aptitude and yum on Linux systems.

The only issue I had when first installing it was that I didn't have Ruby installed on my system (it was an older version of the Mac operating system), and you need to have Ruby to install homebrew. Luckily, Ruby 2.0 ships with OS X El Capitan, Yosemite, Mavericks, and Sierra, and Ruby 1.8.7 ships with Mountain Lion, Lion, and Snow Leopard.

Obviously, not all software packages are `brew`-installation compatible, but it's a nice surprise when you find one that is. One simple command, and it's installed for you—almost as good as a Linux box.

Linux

As with Mac and Windows, you'll need to install the arm-none-eabi-gcc compiler in order to cross-compile for your board. On systems other than Ubuntu, you can do this with a simple

```
sudo apt-get install gcc-arm-none-eabi
```

On Ubuntu, however, you'll need to use a package that's maintained by ARM, `gcc-arm-embedded`. First, remove the built-in package if it's installed on your system with this command:

```
sudo apt-get remove binutils-arm-none-eabi
gcc-arm-none-eabi
```

Then add the correct repo:

```
sudo add-apt-repository ppa:team-gcc-arm-embedded/ppa
```

And then update and install:

```
sudo apt-get update
sudo apt-get install gcc-arm-embedded
```

And finally, you'll need to set your target board with yotta and download the library. Open a new yotta window and enter

```
yotta target bbc-microbit-classic-gcc
```

followed by

```
yotta install lancaster-university/microbit
```

The first time you do this, you may need to log in to your mbed developer account, so make sure you've verified your email address beforehand, since it might need to be verified before you can download the library. Everything you need is now on your machine, and you'll need to do this only once; the next time you open yotta and install the lancaster-university libraries, no login will be needed.

Building an Executable—All Operating Systems

Now that you've got yotta installed and customized, let's use it to build a simple Hello, world application. Create a new directory and cd into it with this:

```
mkdir hello_world
cd hello_world
```

Now initialize a new project with

```
yotta init
```

You'll need to answer some questions during the initialization process. Leave most of them with the default answers (signified by being enclosed in <> brackets) and answer what you need to (Figure 5.7). Make sure, however, that you type **yes** when asked if this project is an executable.

```
● ○ ○                    wolframdonat — activateVE.sh — sh — 80×24
(yotta workspace) ~/microbit_projects $ pwd
/Users/wolframdonat/microbit_projects
(yotta workspace) ~/microbit_projects $ mkdir hello_world
(yotta workspace) ~/microbit_projects $ cd hello_world/
(yotta workspace) ~/microbit_projects/hello_world $ yt init
Enter the module name: <hello-world>
Enter the initial version: <0.0.0>
Short description:
Keywords: <>
Author: Wolf Donat
Repository url:
Homepage:
What is the license for this project (Apache-2.0, ISC, MIT etc.)?  <Apache-2.0>
Is this module an executable? <no> yes
(yotta workspace) ~/microbit_projects/hello_world $ ▮
```

FIGURE 5.7: **Initializing an executable**

Make sure you're targeting the correct board by typing

```
yotta target
```

You should be greeted with this:

```
bbc-microbit-classic-gcc 0.2.3
mbed-gcc 0.1.3
```

If you see

```
bbc-microbit-classic-gcc,* missing
```

make sure you've downloaded the microbit library from lancaster-university before continuing.

Finally, install the libraries you'll need with

```
yotta install lancaster-university/microbit
```

Now that the project has been initialized, you're ready for some simple code. `cd` into the `source/` folder inside your hello_world directory and create a new file called main.cpp. Put the following into main.cpp:

```
#include "MicroBit.h"
MicroBit uBit;
int main()
{
    uBit.init();
    uBit.display.scroll("Hello, world!");
    release_fiber();
}
```

Save it and close it, and then cd back into your main hello_world directory. Enter

```
yotta build
```

and let the project compile. When it's finished, the hex file you need will be inside the build/ directory, at /build/ bbc-microbit-classic-gcc/source, and will be named hello_world-combined.hex. Transfer that hex file to your micro:bit, and be amazed at the "Hello, world!" scrolling across the display! You've just compiled your first C++ program for the micro:bit!

Using a Local Compiler

You may not want to use yotta every time you compile a program for your micro:bit—as I mentioned, it's very much in beta at the moment and is constantly changing. Luckily, if you're running Windows, you can also use Eclipse or Net-Beans as IDEs on your local machine to take care of creating and compiling your micro:bit code. I'll briefly walk you through the process of setting up NetBeans to be your local compiler. If you're using Linux or Mac, sorry, but you're out of luck—getting NetBeans to recognize yotta as a build tool on those platforms is apparently impossible.

Not so on Windows, however: if you haven't done so already, download and install the NetBeans IDE from https://netbeans.org. It requires the Java JDK version 8, so if you don't have that installed, you can get both the JDK and the current version of NetBeans from www.oracle.com/technetwork/java/javase/downloads/index.html.

If you download this IDE, it does not come with the C/C++ plugins installed, so you'll need to install those plugins before you continue.

Once everything is installed, start up NetBeans. You'll need to add c:\yotta (your installation directory) to the system path by editing your system's environmental variables (see Figure 5.6 earlier). Then open the Options window in NetBeans, click the C/C++ tab, and click Add to add a new tool collection. In the window that pops up, enter **C:\yotta\gcc\bin** in the Base Directory field, **Unknown** in Tool Collection Family, and **ARM microbit** in Tool Collection Name (Figure 5.8).

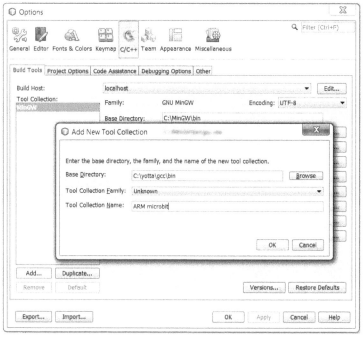

FIGURE 5.8: Setting the base directory

Getting Started with the micro:bit

Then you'll need to manually add the locations of the compilers. Take a look at Figure 5.9 and copy the locations you see in the text boxes.

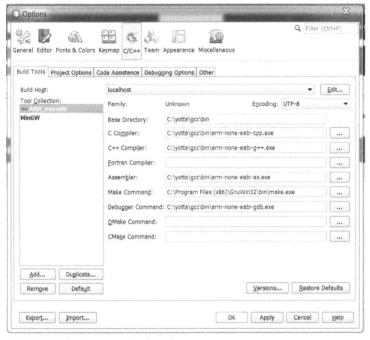

FIGURE 5.9: Setting compiler locations

Now restart the NetBeans interface. When it's open again, go to File > New Project and select a C++ project with existing sources. Then navigate to the `hello_world\build\bbc-microbit-classic-gcc` folder. Set the tool collection to ARM_microbit and click Finish. You'll now be able to edit the `main.cpp` file and clean and compile the project!

I know, that's a lot of work. Don't worry—as I said, we'll use the online compiler in the rest of the book. If you did manage to get all of this working, however, feel free to use your local compiler instead. In the next chapter, we'll take a look at interfacing with the micro:bit using Bluetooth and the radio.

6

Interfacing with the GPIO Pins

Now that you're familiar with the micro:bit board and its onboard parts, we'll look at interfacing with the GPIO pins on the edge of the board. There are several ways to connect to them, the most basic way involving alligator clips and banana plugs.

As you've probably noticed, several of the pins on the edge of the micro:bit are both larger than the other pins and are also connected to holes in the board. This enables you to attach either alligator clips (Figure 6.1) or banana plugs (Figure 6.2) to the board for basic interaction. I won't go into the specifics—readers are more than likely acquainted with

FIGURE 6.1: Alligator clips

these basic interfaces. Just be aware that if you don't have any add-on parts or additional breakout boards, you *can* use these simple parts to do some cool things with your board. All of the functions I discuss in the following sections will work no matter how you connect to the board's GPIO pins. However, I imagine you'll want to add to the basic functionality made possible with clips and plugs, so read on.

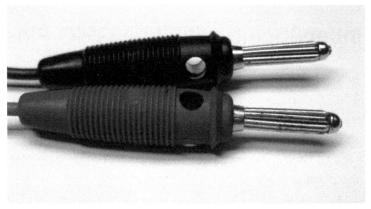

FIGURE 6.2: **Banana plugs**

The GPIO Pins and the Edge Connector Breakout Board

When you've worn out the novelty of basic connectivity, I think it's time to take a look at extending the board's capabilities with some add-on parts. I'd like to introduce you to the *edge connector breakout board*.

I mentioned this board briefly in the beginning of the book. It's available from several online retailers, and should cost you less than $10, including shipping. It has a slot on the side into which you slide your micro:bit board, and it maps all of the "pins" on the edge of the board to actual GPIO headers to which you can attach jumper wires (Figure 6.3). Note the position of the board in the figure—the front of the board faces up, in the same direction as the GPIO pins. It is important to know that for the edge connector board *as well as for the motor controller board*, the orientation of the board is not important when it comes to accessing the pins. The pins are paired; both headers in column 3, for instance, access pin 3 on the board, no matter which direction the board is facing.

It's also worth noting that the rows of pins, although in the same order as the pins on the edge of the board, do

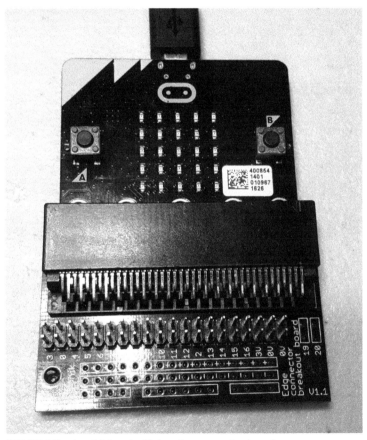

FIGURE 6.3: micro:bit inserted into edge connector board

not follow a numerical sequence—it doesn't start with row 0, followed by row 1, 2, 3, and so on. Rather, the first pair is referred to in code as `pin3`, the second as `pin0`, and so on. See Table 6.1.

TABLE 6.1: GPIO header pin descriptions

ROW LABEL	NAME	DESCRIPTION
20	SDA	I²C SDA—magnetometer/accelerometer (unsoldered)
19	SCL	I²C SCL—magnetometer/accelerometer (unsoldered)
0V	0V	GND
0V	0V	GND
3V	3V	3V
16	DIO	Gen. purpose
15	MOSI	Serial—Master Out/Slave In
14	MISO	Serial—Master In/Slave Out
13	SCK	Serial—Clock
2	PAD2	Gen. purpose
12	DIO	Gen. purpose
11	BTN_B	Button B—goes low on press
10	COL3	LED matrix column 3
9	COL7	LED matrix column 7
8	DIO	Gen. purpose
1	PAD1	Gen. purpose
7	COL8	LED matrix column 8
6	COL9	LED matrix column 9
5	BTN_A	Button A—goes low on press
4	COL2	LED matrix column 2
0	PAD0	Gen. purpose
3	COL1	LED matrix column 1

In front of the header pins, at the very edge of the board, is an area that has been set aside for prototyping, with a positive (3V) rail, a GND rail, and some unconnected pads that can be attached to whatever you like. This makes it easy to connect switches or additional sensors, particularly if you decide to solder some additional header pins to these pads. If you plan on using I^{2}C in your projects at all, I highly recommend soldering some headers to pins 19 and 20. Those two pins are located at the end of the row of headers, rather than inside the prototyping area.

To give a quick introduction to accessing the pins in code, let's go ahead and attach a multimeter to some pins and see what happens in real time as we send commands. Grab a multimeter and set it to either Automatic (if you have a meter that supports it) or DC 1.5V. Attach an M/F jumper wire to pin 0 (the second pin from the left) and another M/F wire to the last GND pin on the right. Attach your multimeter positive and negative leads to those wires. Make sure your micro:bit board is attached to your computer's USB port so you can both power your board and flash programs to it.

Now, on your computer, open up mu, start a new sketch, and press the REPL button to enter an interactive session with your board. Watch your multimeter screen as you type the following:

```
>>> pin0.write_digital(1)
```

Then press Enter/Return. You should get a reading of about 3.23V (Figure 6.4).

Now type:

```
>>> pin0.write_digital(0)
```

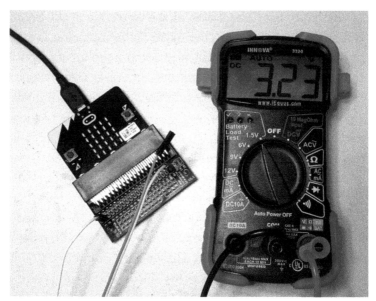

FIGURE 6.4: Turning on pin 0

and press Enter/Return. Your multimeter should read 0V again. If you repeat this experiment—turning on a pin and measuring the voltage—with pin 8, for instance, you should get identical results. You're basically turning those pins on and off by sending digital 1's and 0's to them. Obviously, this can be done within a script as easily as it can be done interactively.

Now hook up your positive jumper wire to pin 5 (the pin for Button A). Your multimeter should read about 3.23V. Press the A button, however, and the voltage should immediately drop to zero, just as it says in the chart. You can repeat the experiment with pin 11 and Button B.

analog_read() and *analog_write()*

Now that you've used `write_digital()`, let's take a look at `read_analog()` and `write_analog()`. These functions allow you to read the value of an analog device, like a potentiometer, for example, or to output an analog value rather than a simple digital 1 or 0. There's an easy way to *display* that value as well—by using an LED hooked up to another pin.

For this experiment, you'll need a potentiometer and an LED. Any color LED will work, so just grab whatever you've got lying around. If you don't have an extra LED lying around, go find/buy/get one, and resolve to never be without a spare LED in your workshop again.

Using jumper wires, connect the pot's GND post (if the pot's shaft is pointing toward the ceiling and the pins are facing you, the GND pin is normally the pin on the left) to one of the micro:bit's GND pins. Connect the pot's input pin (either the middle pin or the one on the right, depending on the pot) to pin 1 on the edge connector board. Connect the other unused pin on the potentiometer to the micro:bit's 3V pin. Finally, connect an LED to the micro:bit's GND pin and pin 2. Everything should look like it does in Figure 6.5.

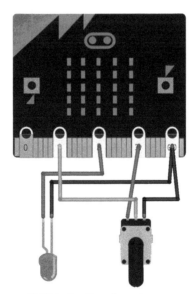

FIGURE 6.5: Diagram of LED/potentiometer setup

Start a new sketch in mu and enter the following code:
Now type:

```
>>> pin0.write_digital(0)
from microbit import *
pin2.write_analog(0)
while True:
    pin2.write_analog(pin1.read_analog())
```

Flash the sketch to your micro:bit.

When you look at your LED's brightness, you may notice that it's not so much getting brighter and dimmer as it is flashing more quickly or more slowly, depending on the position of the pot. That's because the micro:bit is actually using something called pulse width modulation (PWM) to light the LED and change its brightness level.

Pulse width modulation is a way of producing analog effects using digital means. Many electronic devices you may be familiar with (such as the micro:bit, the Arduino, and the Raspberry Pi) are unable to send varying degrees of voltage out through a GPIO pin—they can only turn it on or off. In order to simulate various voltage levels, therefore, they vary the *duty cycle* of the pin using a square wave.

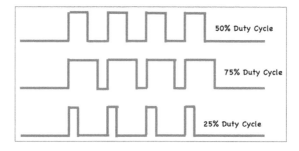

In other words, PWM involves turning a digital pin on and off very quickly. The more time the pin is on over the course of a second, the longer the duty cycle and the higher the simulated voltage. When you see the LED flashing, you're actually seeing the duty cycle pulses the board is sending through pin 2 to the LED.

PWM isn't just used to light LEDs; one of its other more common uses is to control servomotors. Varying duty cycles are used to position the servo at different points along its range of travel (or to control its speed and direction, in the case of a continuous-motion servo). Luckily, most electronic devices that use PWM use libraries (Arduino's `Servo.h` library and Raspberry Pi's `PiGPIO` library, for example) that take care of the actual duty-cycle programming for us behind the scenes.

The micro:bit is no exception.

Assuming you've got everything wired correctly, your LED should immediately light up, though it may be rather dim depending on your potentiometer's position. Play around with the pot's setting, and you should see the LED change brightness depending on the position of the pot.

Looking at the script, it should be pretty self-evident what's going on here. We start writing a 0 to pin 2 to make sure it's completely off. Then we read the value of the potentiometer (or the voltage it's passing, to be technically correct) and send that voltage to the LED. The more voltage the pot is passing, the brighter the LED.

If you're interested in the actual values being emitted by the potentiometer, you can try the following script:

```
from microbit import *
while True:
    text = str(pin1.read_analog())
    display.scroll(text)
    sleep(500)
```

For this script, remove the LED from the circuit. Connect the pot's ground pin to the micro:bit's GND pin, the pot's input pin to pin 1, and the pot's other pin to the micro:bit's 3V pin (Figure 6.6.) When you flash the script to your board, it will display the pot's current value every half second. The micro:bit has a 10-bit analog-to-digital converter on board, so you should see values ranging from around zero all the way to 1023 at the other end of the pot's travel.

Now that we've looked at `read_analog()`, `write_analog()`, and `write_digital()`, there's one more function in that group we should investigate—`read_digital()`. Just as you'd suspect, this function is what you'll use to read from sensors such as switches that send a simple ON/OFF signal to the board. If you happen to have an old switch lying around, it's a perfect way to test `read_digital()`. If not, you'll just have to trust me on this one. Hook up a simple single-pole,

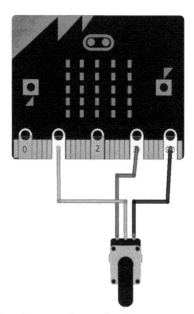

FIGURE 6.6: Reading the potentiometer's value

single-throw (SPST) toggle switch to pin 16 and one of the 3V pins. With the switch open, typing

```
>>> pin16.read_digital()
```

on the mu REPL command line should return 0. Running the same command with the switch closed will return a 1. Quite simple. In the same vein, the `is_touched()` function is interesting. You don't need a switch to play with this one, as your body acts like a switch. Connect one of your jumper wires to pin 1 and the other to a GND pin. If you hold both wires in your hand,

```
>>> pin1.is_touched()
```

will return a 1, whereas leaving one wire untouched will return a 0. Connecting the pin to ground (through your body) sets the return value to 1.

I²C

One of the things that hobbyists may find attractive about the micro:bit is its I²C capabilities. Pins 19 and 20, though not immediately available without soldering some headers to them, allow you to control and read from I²C devices, which can be *very* helpful when you're attaching various sensors and devices. Having a preprogrammed way of interacting with large numbers of add-ons can greatly simplify your life.

To use the I²C interface with the micro:bit's breakout board, you'll need to solder some headers to pins 19 and 20. As you remember from Table 6.1, pin 19 is the clock line (SCL) and pin 20 is the data line (SDA). You can daisy-chain your devices just as you normally would, and then the functions `i2c.write()` and `i2c.read()` allow you to communicate with and control them from your board. You don't need to specify a pin when you call these functions, since pin 20 is hard-coded to read and write I²C data, and the micro:bit's firmware automatically takes care of the timing for you with pin 19—the SCL (clock signal) pin.

If you've been around small embedded electronics like the Arduino and the Raspberry Pi for a while, you're probably familiar with I²C. If, however, you're new to the game, you might be wondering just what it is.

I²C, pronounced "Eye Two See," "Eye Eye See," or sometimes "Eye Squared See," stands for Inter-Integrated Circuit, and was developed by Philips Electronics in the early '80s. It's a simple two-line system that can be used as a bus for a large number of devices. It typically has one *master* device and many *slaves*, each of which has a unique address of 7 to 10 bits. The clock signal synchronizes all the devices and allows them to communicate without stepping all over each other.

In a typical scenario, the master device decides it wants some information from a device—let's call it device X. On the clock tick, the master sends out a signal on the data line: "Device X, please send me your data." Devices A, B, C, and D are listening as well, but as soon as they hear that the call is for device X, they stop paying attention. Device X, on the other hand, perks up its ears. On the next clock tick it responds with an "OK," and then proceeds to send the data. In another scenario, the master device may have a command for device C; in this case it advertises device C's address first, getting its attention, and then sends it the command. While I²C is a very simple protocol, it's also very useful. Many, many devices use it, and its low power usage makes it an attractive option for devices like the micro:bit.

This brings up another helpful hint for you: if you're not sure what functions are available for each pin (not all pins support `read_analog()` and `write_analog()`, for instance) you can type the pin number in your mu REPL window, followed by a period and the Tab key, and you'll see a list of all functions that are available to that pin (Figure 6.7.)

```
MicroPython v1.7-9-gbe020eb on 2016-04-18; micro:bit with nRF51822
Type "help()" for more information.
>>> pin6.
write_digital    read_digital     write_analog
set_analog_period                 set_analog_period_microseconds
get_analog_period_microseconds
>>> pin6.
```

FIGURE 6.7: List of available functions for pin 6

If you type **i2c.write(** (no closing parenthesis) and press Tab, you'll see all the possible parameters for the `i2c_write()` function. You'll notice that pin numbers are allowed, but that's not for reading and writing I²C values; rather, those parameters are available should you want to write the value of `pin2.read_analog()`, for example.

The Motor Driver Board

Now that you're familiar with ways to access your board's GPIO pins, let's talk about possibly the coolest way of interacting with your board's pins: with a motor controller board, which I mentioned in the first chapter. We'll be using that board for the micro:bot in a later chapter, so let's see how we can interact with it. (See what I did there? The *micro:bot* is a play on words, and that's also what we literary types refer to as *foreshadowing*. Stay tuned!)

The motor controller board has a slot into which you'll plug your micro:bit, a row of pads on the edge corresponding to all of the micro:bit's pins (so you can attach more stuff even when the board's plugged in), and seven block terminals.

These terminals allow you to connect an external power supply (essential for driving motors, since the micro:bit can't source much current), and connect to two different motors (such as a left drive wheel and a right drive wheel). You can also connect two separate input devices that will be hardwired to pins 1 and 2, and you can also connect an external device to the button_a and button_b inputs to be read by the micro:bit.

To use the motor controller board, plug your micro:bit into the slot. The board is designed so that the micro:bit board may be facing either way; however, in order to use the extra row of headers on the edge of the board, the LED array will need to face that edge (Figure 6.5). For our testing purposes here, it doesn't matter which way you plug it in.

Now plug in a power supply. The screen printing on the controller board suggests between 4.5V and 6V, so try a pack of 3 or 4 AA batteries. The maximum recommended voltage is 6 volts, so please don't blame me if you plug in a car battery and either your motor controller board or your micro:bit releases a puff of smoke!

Finally, plug in two jumper wires to the MOTOR1 terminal block. Connect these to your multimeter, using either the AUTO setting or the 1.5V DC setting as we did before. Your final setup should look like Figure 6.8.

FIGURE 6.8: Testing setup for motor control board

As with most motor controller boards, we control the motors by alternating the signals sent to the motor terminals. MOTOR1 is controlled with pins 8 and 12 on the micro:bit, and MOTOR2 is controlled with pins 0 and 16. Table 6.2 shows the motor behavior associated with different pin values; remember that Forward and Reverse are relative and depend on how the motor is connected in the first place.

TABLE 6.2: Pin values and motor directions

P8	P12	MOTOR1	P0	P16	MOTOR2
0	0	coast	0	0	coast
1	0	forward	1	1	forward
1	1	brake	1	1	brake
0	1	reverse	0	1	reverse

To illustrate this, our multimeter will show us voltages as we change pin values around. In your mu window, start a new Python script and enter the following:

```
from microbit import *
while True:
    # Reverse
    pin12.write_digital(1)
    pin8.write_digital(0)
    sleep(2000)
    # Coast
    pin12.write_digital(0)
    sleep(2000)
    # Forward
    pin8.write_digital(1)
    sleep(2000)
    # Brake
    pin12.write_digital(1)
    sleep(2000)
```

Flash this code to your micro:bit and watch the display on your multimeter. It should start switching between your connected voltage, zero, and a negative voltage, about two seconds apart. I'm using 5 AA batteries in the setup shown in Figure 6.5, so my meter shows 7.2V, 0V, –7.2V, 0V, and so on. When the voltage is positive, a connected motor will be spinning in one direction, and when it's negative the motor will spin in the opposite direction.

As I've mentioned before, two of the terminal blocks on the board give you access to the button pins; in other words, sending voltage to those blocks has the same effect as pushing the buttons on the micro:bit board. To illustrate that, keep the same setup described earlier, and connect an additional set of jumper wires to the terminal block for Button A. You don't have to connect any voltage source—pressing the buttons just sends them to ground.

In a new mu script, enter the following:

```
from microbit import *
while True:
    pin12.write_digital(0)
    pin8.write_digital(0)
    if button_a.is_pressed():
        pin12.write_digital(1)
        pin8.write_digital(0)
    else:
        pin12.write_digital(0)
        pin8.write_digital(1)
```

Flash this code to your board. When it's running, at first your multimeter should be displaying a negative voltage. When you touch the two jumper wires together, however, the voltage should immediately become positive. To double-check that it's the same as pressing the button, try pressing the actual Button A on the micro:bit—the results should be the same.

Finally, to try out the input terminal blocks, connect the same two jumper wires to the INPUT2 block and flash this code to your micro:bit:

```
from microbit import *
while True:
    pin12.write_digital(0
    pin8.write_digital(0)
    if pin2.is_touched():
        pin12.write_digital(1)
        pin8.write_digital(0)
```

Now your multimeter should read around 0V unless you touch the INPUT2 wires together. At that point, you should see a negative value on the meter, which then goes away when you release the wires.

You can also substitute if `pin2.read_digital() == 1:` for the if statement in your code, but I've found that unless you have fresh batteries that deliver at *least* 3V, the pin will never read 1. For that reason, I'm not particularly fond of the `read_digital()` function and would rather use either `is_touched()` or `read_analog()` and use the value of a pot to determine an input as a parameter value.

So that's an introduction to accessing the GPIO pins on the micro:bit. There are additional functions that I didn't go into here, but this should give you a basic idea of what's possible, with either the edge connector board or the motor controller board. In the next chapter I'll show you how to use Bluetooth and the onboard radio to communicate with your board, all in preparation for the projects in Chapter 8.

7

Using Bluetooth

Now that you're familiar with connecting to the micro:bit using hardware and wires, it's time to take a look at connecting to the board using the Bluetooth stack. After all, one of the big selling points of the micro:bit is its BLE (Bluetooth Low-Energy) connectivity, which is used in many different IoT applications and devices.

What Is Bluetooth?

Before we get started, I feel I should warn you that Bluetooth connections in general can be somewhat finicky, and communicating with the micro:bit is no different. Things don't always work the first time, and the micro:bit is still relatively new and the software and firmware are still being updated. Just don't get too frustrated if things seem a little wonky, and I'll try to walk you through it as gently and completely as possible.

So what exactly is Bluetooth? You've most likely used it with your smartphone to connect to external devices like speakers and headphones, or your car radio, or even to send files to another smartphone. Or you may have used it on your computer or laptop—perhaps to use a wireless mouse or keyboard. It's been around longer than you may think (since 1994, as a matter of fact) and version 5.0 is now available, which, according to the Bluetooth Special Interest Group, quadruples range, doubles speed, and increases data broadcasting capacity by up to 800 percent. If you have a moment, I recommend visiting www.bluetooth.com/about-us/our-history, as it's quite an enlightening read.

Bluetooth is a wireless technology standard originally designed to be a wireless substitute for RS-232 (serial) communication, developed by Ericsson Telecommunications. Ericsson was mainly interested in two things: developing wireless headsets, and allowing mobile phones to communicate with computers. Devices using Bluetooth communicate at frequencies between 2400 and 2483.5 MHz. Within that spectrum there are seventy-nine designated channels over which standard devices can communicate; BLE devices like the micro:bit have only forty channels to choose from.

Bluetooth differs from WiFi in two main ways: power and connectivity. WiFi was designed to replace Ethernet and coaxial cables comprising a network, and is much more powerful in terms of broadcast strength. WiFi is also normally set up in a client-server mode, where most networked devices communicate with each other via a central node, such as a router or a switch. Bluetooth, on the other hand, tends to be a symmetrical setup, where devices communicate directly with each other.

Bluetooth LE, the specification used by the micro:bit and many other IoT devices, was released as Bluetooth 4.0 in 2010. It's intended to require considerably less power and cost less than classic Bluetooth but to be similar in terms of communication range. IoT devices that use BLE can go for extended periods of time—often months or even a year or more—on the same small battery.

Programming Bluetooth on the micro:bit

When it comes to using Bluetooth on your micro:bit, you'll most likely end up using C++ and your mbed developer account, because as you'll remember from Chapter 5, "The mbed Operating System," many of the other online programming environments don't get you close enough to the hardware to program the Bluetooth stack directly. MicroPython doesn't offer Bluetooth connectivity at all, and some of the others offer only a limited range of options and functionality. C++ and mbed are currently the best ways to experiment with Bluetooth (though I expect that may change as the development environment matures).

Another thing you'll probably need in order to conduct pairing experiments is an app for your smartphone that can read and write to the micro:bit. Unfortunately, as it stands right now, there is a serious lack of iPhone apps that are designed for Bluetooth development work with the micro:bit. I've done my best to research apps for both platforms, but it seems that Android is the way to go if you're looking to do any serious work with Bluetooth on your board. The best cross-platform app I've found so far is called nRF Connect for Mobile. It is free and is available for both Android and iPhones—just search for it in your respective app store and download it to your device. It's not particularly user-friendly and is quite technical, but it is handy for showing that you're connected to your board, it gives you some basic functionality, and it works on both types of smartphones. If you have an Android phone, however, I highly recommend picking up Martin Woolley's micro:bit Blue app, also available for free at the Google Play Store. It allows you to read and write to the micro:bit device over Bluetooth in a graphic environment, as I'll show you in a moment.

When we go on to Chapter 8, I'll discuss serial communication over Bluetooth, which can be done either with an iPhone or an Android, or even a computer with a Bluetooth adapter installed. I also highly recommend the official micro:bit app (Figure 7.1), available for both the iPhone and Android devices for free at their respective app stores (just search for *micro:bit*). The official app is generally pretty helpful and is designed for the novice user. It can get finicky at times as well, but it will allow you to pair your micro:bit, write code directly on your phone, and then flash that code to your board over Bluetooth.

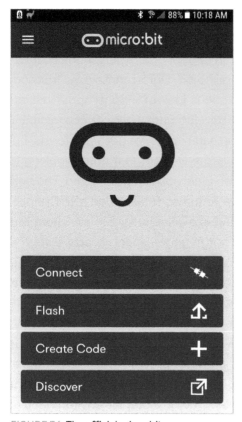

FIGURE 7.1: The official micro:bit app

Getting Started with the micro:bit

To get started, however, let's start with mbed. On your computer, log into your mbed account and bring up the compiler workspace. First, let's make sure you're connected and have the correct libraries available. Click New at the top left of your IDE menu bar. In the pop-up menu, make sure that "An example of how to use the micro:bit DAL's abstraction" is selected (Figure 7.2). In this context, DAL stands for Device Abstraction Layer, and is an abstraction of the hardware on the micro:bit that has been written by the good folks at Lancaster University. It creates the `MicroBit` class and allows you to access various parts and functions of the device by using methods and variables that have been written into the class.

Name the program whatever you like (*microbit-test* is a good name) and click OK.

Now, you'll compile it. Make sure that microbit-test is selected in your list of programs, then click Compile in the menu bar. It may take a few minutes to compile, but then you'll get a download prompt for a hex file. Download the file and flash it to your micro:bit, and you should be rewarded with "HELLO WORLD! :)" scrolling across your display.

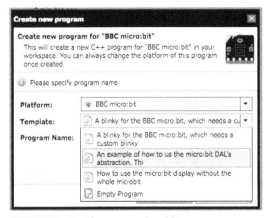

FIGURE 7.2: Starting a new micro:bit program

Great! Now you know that your compiler is set correctly and you can flash the resulting files to your board. In order to work with Bluetooth, you should import an already-working program into your compiler that you can play around with and change. In a new tab, point your browser to https:// developer.mbed.org/teams/BBC/code/microbit-samples/, where you'll find a sampling of some of the possibilities for your board. Click the Import Into Compiler button on the right, and import the program into your compiler workspace.

Once the program is imported, expand it in the Program Workspace on the left, and then expand the `source` folder. Right-click on the `HelloWorld.cpp` file and choose Clone (Figure 7.3). Call it `BLE.cpp`.

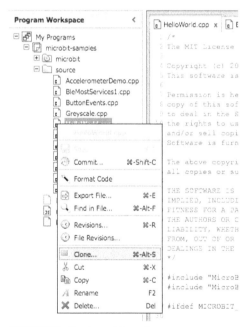

FIGURE 7.3: Cloning `HelloWorld.cpp`

Now open the `MicroBitSamples.h` file. Line 41 or so should read as follows:

```
#define MICROBIT_SAMPLE_HELLO_WORLD
```

Change that line to read

```
#define MICROBIT_BLE_SERVICES
```

Then, in your `BLE.cpp` file, change line 29 to read

```
#ifdef MICROBIT_BLE_SERVICES
```

Now scroll down a bit (to line 41 or so) and add the following lines to your `main()` function:

```
// Gives Bluetooth access to the various micro:bit
// functions
new MicroBitLEDService (*uBit.ble, uBit.display);
new MicroBitTemperatureService(*uBit.ble, uBit.
thermometer);
new MicroBitAccelerometerService(*uBit.ble, uBit.
accelerometer);
new MicroBitMagnetometerService(*uBit.ble, uBit.
compass);
new MicroBitButtonService(*uBit.ble);
new MicroBitIOPinService(*uBit.ble, uBit.io);
```

These lines are initializing the Bluetooth on your micro:bit and making the various services (`display`, `thermometer`, etc.) available to the Bluetooth stack, which makes them visible to other devices.

Now, you'd hope that would be all you'd need to do and you could go happily pair your micro:bit to your phone. Unfortunately, there's another problem: Bluetooth services use a *lot* of memory. Since your board doesn't have much to begin with, it turns out that this program, as written, will take up all the memory on your board and the pairing will fail. So you need to change a few things so your program isn't such a memory hog.

Back in your compiler window, expand the `microbit/`
`microbit-dal/inc/core` folder and click on the `MicroBitConfig.h`
file (Figure 7.4).

Scroll down to line 236, and change the `1` at the end of
the line to a `0`. Likewise, on line 243, change that `1` to a `0`. Here
you're turning off the device firmware update and event services, since you don't need them in this application (you're
just testing the Bluetooth, after all). When you're done, line
236 should read

```
#define MICROBIT_BLE_DFU_SERVICE         0
```

and line 243 should read

```
#define MICROBIT_BLE_EVENT_SERVICE       0
```

Next, in the same file, scroll up to line 103, and edit it to read

```
#define MICROBIT_SD_GATT_TABLE_SIZE
0x700
```

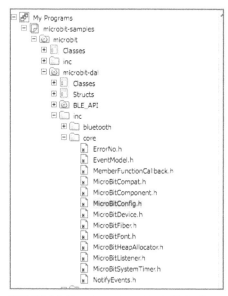

FIGURE 7.4: The `MicroBitConfig.h` file in the source tree

This GATT_TABLE is where the micro:bit keeps track of Bluetooth services. By changing the value to 0x700, we've maximized the size of that table.

Finally, scroll up a bit further, and change the value at the end of line 89 to 0.50 rather than its current 0.75, so that it reads

```
#define MICROBIT_NESTED_HEAP_SIZE       0.50
```

This slightly decreases the amount of memory on the heap available and gives you a bit more for your program.

When you've finished, click the small arrow inside the Compile button and click Compile All (Figure 7.5). After a bit you'll be rewarded with a hex file that you can download and flash onto your micro:bit.

FIGURE 7.5: Compiling the program

When it's flashed, the first thing you'll see is the "Draw a circle" message. That's because you included the compass service in your Bluetooth tests, and the compass needs to be calibrated every time you flash the device. Draw the circle to calibrate the compass, and then turn on your phone's Bluetooth and open the Bluetooth settings.

Now you'll pair the micro:bit to your phone. To do this, on your micro:bit press and hold the A and B buttons, then press the reset button on the other side of the board. Hold down the reset button for a second or two, and then release it. Then release the A and B buttons. It all sounds a bit more complicated than it is; in short, you just press and hold A and

B, press and release reset, and then release A and B. You'll see "PAIRING MODE!" scroll by on the display, and then a pattern will appear on the LED matrix (Figure 7.6).

When you see this pattern, use your phone to try to connect to the micro:bit. The board should immediately display an arrow pointing to the A button, and your phone should ask for a six-digit code. This is a code randomly generated by the micro:bit that needs to be entered into the phone to complete the pairing process. Press the A button on the board, and six digits will flash, one after the other. Enter these numbers into your phone, and you should be greeted by a check mark on the device.

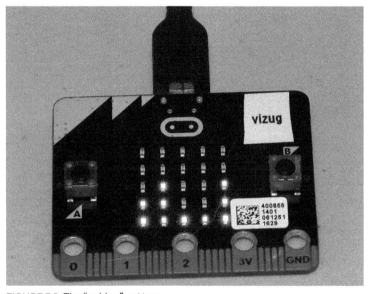

FIGURE 7.6: The "pairing" pattern

NOTE Your phone may display a message that pairing was unsuccessful; if you got a check mark on the device, you can safely ignore that message—it's paired.

If you're using the official micro:bit app, the app will ask you to duplicate the pattern first, and then will go through the six-digit pairing process. The end result is the same.

Now that you're connected, open up the nRF Connect app. On the Scanner tab, find your micro:bit in the list of available devices and click Connect. After the phone connects to your board, you should see a long list of client attributes and services that are being read off your board. The services you enabled in your BLE.cpp file are listed here as Unknown Services. Once you're connected, you are able to read and write, where applicable, to these services, using your phone. Perhaps the easiest way to see this is to expand the top Generic Access service and read the device name from your board by clicking the down-facing arrow. You should see something like Figure 7.7, with the device name filled in.

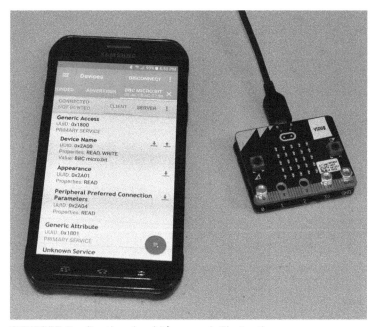

FIGURE 7.7: Reading the micro:bit's name via Bluetooth

You can also click the up-facing arrow (which is short-hand for *write*) and write a new name to the board. It won't last through a reboot, but it's a simple way of making sure everything is working.

Now, if you happen to have an Android phone and have downloaded the micro:bit Blue app from Martin Wooley I mentioned earlier, you can see a bit more information. Start the app and select your micro:bit from the list of paired devices. You may need to click "Find paired BBC micro-bits" at the bottom. Once you've connected to your board, you'll see a list of available services, as shown in Figure 7.8.

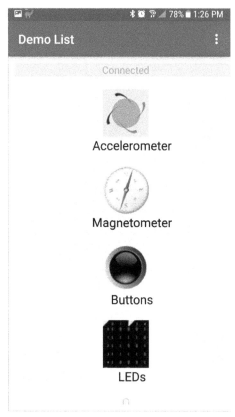

FIGURE 7.8: Available Bluetooth services

Getting Started with the micro:bit

Play around with the services we enabled, like the accelerometer, the buttons (Figure 7.9), the thermometer (Figure 7.10), and—perhaps the most fun—the display LEDs (Figure 7.11).

Unfortunately, there is no similar app available yet for the iPhone, so for now you'll have to be satisfied with checking that the Bluetooth services exist with the nRF app. In the next chapter, however, we'll discuss using serial communication over Bluetooth to send messages back and forth, and that communication can happen no matter what platform you happen to be using.

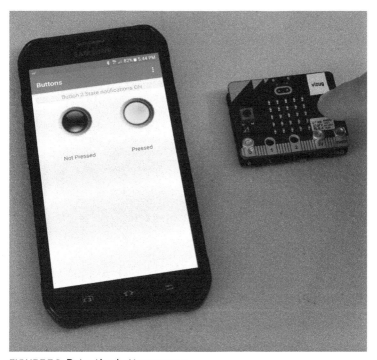

FIGURE 7.9: Detecting button presses

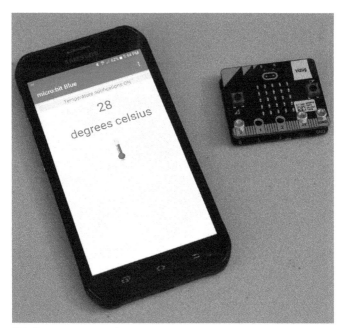

FIGURE 7.10: Reading from the thermometer

FIGURE 7.11: Writing to the display

Third-Party Apps

All of this may seem a bit complicated, and as I mentioned earlier it's because some sort of program needs to be running on your smartphone in order to truly communicate with the micro:bit. You can program your board to send and receive Bluetooth messages, but you still need a device to handle the other side of the communication, and programming an external device like a smartphone to handle the other end of the messages in a GUI like an app is obviously outside of the scope of this book.

Luckily, thanks to the rise in popularity of smartphone apps , there are other options when it comes to building one for your phone. There are some simple plug-n-play apps available that can work with the micro:bit, and I'm sure that more will arise as the micro:bit's popularity increases. Probably the most successful so far is the Evothings environment. Evothings uses a somewhat unique model: there are actually three programs working simultaneously. The first, obviously, is your micro:bit code. A second code runs on your phone and allows you to program a GUI using drag-and-drop technology. It also allows you to interact with an external device like a micro:bit. Finally, a third program, Evothings Studio, runs on your desktop or laptop computer (Figure 7.12) and communicates with your phone via Evothings' cloud servers. These cloud servers act as a middleman between your computer and your phone. It sounds complicated, but it seems to work pretty well. If you're interested in playing with Evothings, you'll need the mobile app for your OS, and the Evothings Studio program for your computer, available at https://evothings.com.

Kivy is another cross-platform Python development tool, designed to make coding apps as painless as possible, but unfortunately Kivy's Bluetooth support is almost nonexistent, so it won't work for our purposes. Other apps exist as well, but it's hit and miss when it comes to their success with working with the micro:bit.

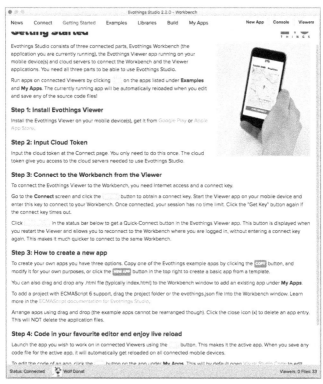

FIGURE 7.12: Evothings Studio

As I said earlier, in the next chapter we'll break things down further and make them even simpler, and communicate with the micro:bit via serial communication over Bluetooth. This approach is not particularly elegant, but it will allow us to communicate with and control a small mobile robot, and Bluetooth serial communication applications are available for both platforms.

8

Serial Bluetooth Communication and the micro:bot

Now that we've looked at some Bluetooth setup and configuration, let's explore another way to interact with your micro:bit over Bluetooth—with the UART serial interface. A UART (universal asynchronous receiver/transmitter) is a form of communication via hardware, over a serial connection, in which the data format and transmission speeds are configurable.

This form of communication can sometimes be difficult to set up, but once it's working it's the simplest way of sending and receiving messages to your micro:bit board. Since you're going to want to control your little mobile robot via your board with Bluetooth from your phone, it makes sense to learn about this way of communicating.

UART and Bluetooth

The serial UART Bluetooth interface is not enabled by default, so you'll need to turn it on and use a few lines of code to configure it. Again, this is beyond what MicroPython, Blocks, and Code Kingdom are capable of, so you'll need to open up your mbed compiler window again.

Let's start by cloning the `microbit-samples` code again like we did in the previous chapter. Right-click on the `microbit-samples` folder and choose Clone. Name the new folder whatever you like, but as this is testing our UART communications, let's name it something appropriate. I named mine UART.

Once the cloning process has completed, you have several things to do. You'll have to change the `MicroBitSamples.h` file to `#define` a new variable. You'll have to write a new `.cpp` file that compiles and does what you want it to once the new variable has been defined. And finally (and as it turns out, *most importantly*), you'll need to update your `microbit-dal` folder.

As it turns out, the `microbit-dal` folder that's included with the `microbit-samples` code is slightly outdated. It works for most purposes, but the included files don't declare some important variables—namely `MICROBIT_BLE_EVT_CONNECTED` and `MICROBIT_BLE_EVT_DISCONNECTED`. Those variables are used to call functions when particular events (Bluetooth connections and disconnections, as you probably guessed) are detected. In order for these events to be detected correctly (and for your compilations to complete without errors), you must obtain a newer version of `microbit-dal` and replace the one currently residing in your UART folder. So first, in your current UART folder, right-click the `microbit-dal` directory (`UART/microbit/microbit-dal`) and click Delete.

Make sure you're still signed in to your mbed account. Point your browser to https://developer.mbed.org/teams/Lancaster-University/code/microbit-dal/rev/493daf8966fd. Just like you did for the microbit-samples application, click the orange

Import Into Compiler button on the right side of the page (Figure 8.1).

FIGURE 8.1: Importing a new version of `microbit-dal`

The Import Library window will open; this is where you tell the compiler where to place the new library. Make sure Library is selected, and make the target path **/UART/microbit** (Figure 8.2). Then click Import.

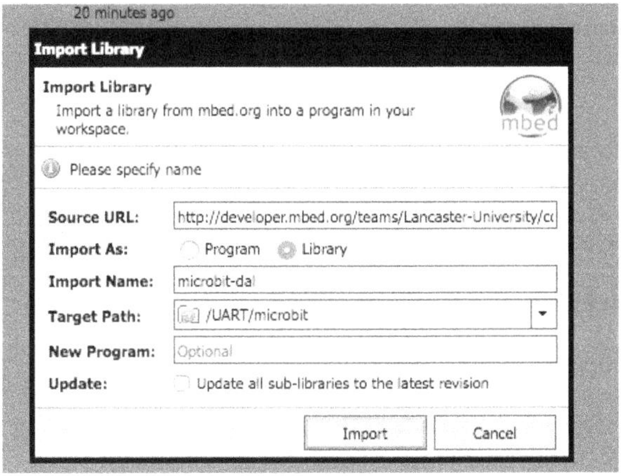

FIGURE 8.2: Placing the new `microbit-dal` library

Now that's done, you can start on the actual code. Open the `MicroBitSamples.h` file (`/UART/source/MicroBitSamples.h`). Make sure all of the `#defines` between lines 38 and 50 are commented out (with a double `//` in front). Then, before all of them (line 40 or so), add the line

```
#define     MICROBIT_UART
```

and save the file.

Now, right-click on the `source` folder and choose New File. Name the new file **UART.cpp**. Then write the following in the file (we'll go over this in a moment):

```
#include "MicroBit.h"
#include "MicroBitSamples.h"
#include "MicroBitUARTService.h"
#ifdef MICROBIT_UART
MicroBit uBit;
MicroBitUARTService *uart;
int connected = 0;
void onConnected(MicroBitEvent e)
{
    uBit.display.scroll("C");
    connected = 1;
    ManagedString eom(":");
    while (1)
    {
        ManagedString msg = uart->readUntil(eom);
        uBit.display.scroll(msg);
    }
}
void onDisconnected(MicroBitEvent e)
{
    uBit.display.scroll("D");
    connected = 0;
}
void onButtonA(MicroBitEvent e)
{
    if (connected == 0)
    {
        uBit.display.scroll("NC");
        return;
    }
    uart->send("YES\n");
```

```
    uBit.display.scroll("YES");
}
void onButtonB(MicroBitEvent e)
{
    if (connected == 0)
    {
        uBit.display.scroll("NC");
        return;
    }
    uart->send("NO\n");
    uBit.display.scroll("NO");
}
void onButtonAB(MicroBitEvent e)
{
    if (connected == 0)
    {
        uBit.display.scroll("NC");
        return;
    }
    uart->send("GOT IT!\n");
    uBit.display.scroll("GOT IT!");
}
int main()
{
    uBit.init();
    uBit.messageBus.listen(MICROBIT_ID_BLE, MICROBIT_
BLE_EVT_CONNECTED, onConnected);
    uBit.messageBus.listen(MICROBIT_ID_BLE, MICROBIT_
BLE_EVT_DISCONNECTED, onDisconnected);
    uBit.messageBus.listen(MICROBIT_ID_BUTTON_A,
MICROBIT_BUTTON_EVT_CLICK, onButtonA);
    uBit.messageBus.listen(MICROBIT_ID_BUTTON_B,
MICROBIT_BUTTON_EVT_CLICK, onButtonB);
    uBit.messageBus.listen(MICROBIT_ID_BUTTON_AB,
MICROBIT_BUTTON_EVT_CLICK, onButtonAB);
    uart = new MicroBitUARTService(*uBit.ble, 32, 32);
    uBit.display.scroll("AVM");
    release_fiber();
}
#endif
```

So let's take a look at this code for a moment (you can download the file at https://github.com/wdonat/microbit-code/blob/master/chapter8/UART.cpp). The first two lines you've seen before, and the third line, #include "MicroBitUARTService.h", is necessary to start the UART service, since it's disabled by default.

```
MicroBit uBit;
MicroBitUARTService *uart;
```

Then we initialize the `MicroBit` object and create a pointer to a `MicroBitUARTService`, `uart`. This explains the `uart->` lines later in the program; `uart` is a special memory location that points to a collection of functions and variables that all work together, and those functions and variables are accessed using the arrow (`->`) notation. If you were creating a class and accessing its functions and variables directly, rather than via a pointer, you would use a dot (`.`) notation instead.

The next two functions, `onConnected()` and `onDisconnected()`, dictate what happens when the micro:bit detects an event—namely, `MICROBIT_BLE_EVT_CONNECTED` and `MICROBIT_BLE_EVT_DISCONNECTED` that we see in the main function. (You can see why we needed the newer version of the `microbit-dal` directory—that's where those two variables are declared and defined.) In the main function, you can see that we're listening to the `messageBus` on the micro:bit, and calling those functions when we detect the correct event.

There are two important lines in the `onConnected()` function regarding the end-of-message (`eom`) variable. You can see that it's a `ManagedString`, that its value is `:`, and that we call the `uart->readUntil()` function with it. What this means is that we're using the colon character (`:`) as our end-of-message marker, and the micro:bit will read (and display) the message we send until it gets a colon.

The other three functions dictate what happens when we detect a `MICROBIT_BUTTON_EVT_CLICK` on the `messageBus`—either Button A, Button B, or both. Obviously, for testing purposes, we're not doing anything spectacular—just displaying and returning some text.

Open the `MicroBitConfig.h` file (`UART/microbit/microbit-dal/inc/core/MicroBitConfig.h`). Again, you'll need to do a little memory management, since the Bluetooth is such a memory hog. On line 90, set the heap size to 0.50:

```
#define MICROBIT_NESTED_HEAP_SIZE        0.50
```

A little further down on line 104, make the GATT table a bit smaller:

```
#define MICROBIT_SD_GATT_TABLE_SIZE    0x300
```

If you'll recall from Chapter 7, "Using Bluetooth," the GATT table is where the micro:bit keeps track of Bluetooth services. Since you're not using all of the available services, you can make it smaller, which has the desirable effect of freeing up some memory for you.

And finally, turn off some services. On line 237, disable the DFU service (this disables on-air programming, which we're not doing anyhow):

```
#define MICROBIT_BLE_DFU_SERVICE    0
```

and on line 244, disable event services, since they're not needed here:

```
#define MICROBIT_BLE_EVENT_SERVICES    0
```

While we're in this file, there are two tweaks I make that seem to make Bluetooth development a bit easier. These are completely optional, and your program will work whether or not you try these tweaks. The first thing I do is disable security to make pairing a bit easier. On line 190, set it to 1:

```
#define MICROBIT_BLE_OPEN    1
```

Then set the no-security transmission power on line 198:

```
#define MICROBIT_BLE_DEFAULT_TX_POWER    6
```

Finally, I disable the whitelist on line 216 to make pairing easier:

```
#define MICROBIT_BLE_WHITELIST    0
```

As I said, these last tweaks are optional, but if you're having trouble pairing it might be worth giving these a try.

When you've finished making all of these tweaks, it's time to compile and save the hex file. Make sure everything has been saved, and then click the arrow on the Compile button and select Compile All. When it's finished compiling, save the hex file (which should be named UART_NRF51_MICROBIT.hex) to your computer.

You can flash it to your micro:bit right away if you like, but you'll need something else before you can test it—an app for your smartphone that will allow you to send and receive serial commands over Bluetooth. I don't have a recommendation here; there are multitudes of possibilities for each platform, but choose a free one. The important thing is that it's for serial communications, and it's for both Bluetooth and Bluetooth LE devices (not all apps work with BLE devices). I'm using Serial Bluetooth Terminal by Kai Morich on my Android.

Once you've selected an app, install it, pair your phone (or computer) with the micro:bit, and open the application. Make sure the two devices are paired; you may need to tell the particular app you're using to connect to the micro:bit, though you've already paired them. You may need to press the Reset button on your micro:bit as well. Once they're paired, on your phone or laptop, type **a**, follow it with the colon (**:**), and press SEND.

You should immediately see "a" scroll across the micro:bit's display. Pretty cool, eh? Try typing a word next, like **spam**, and follow it with a colon. Same result, right?

Okay, now press the B button on your micro:bit. You should see "NO" scroll across the display, *and* you should also see "NO" appear in the terminal on your smartphone or computer (Figure 8.3).

FIGURE 8.3: Serial communication via Bluetooth

You've done it! You're now successfully communicating back and forth with your micro:bit! As you can probably imagine, it's only a short step now to communicating with (and controlling) a small mobile robot.

I feel I should preface this section by letting you know that there is at least one other mobile robot out there called the micro:bot, so this is in no way trademarked, and if you can come up with a better name for it, that'd be awesome.

I should also state here that I did not make the platform on which our robot is based. I do tend to build a lot of robots from scratch, but at some point I realized that I was essentially doing the same thing over and over again. So, in the interest of speeding up the process of creating mobile robots using different parts, I invested in a bot development platform. I chose the Turtle, a 2WD mobile platform from DF Robot (https://www.dfrobot.com/product-65.html—Figure 8.4).

FIGURE 8.4: The Turtle platform

I do not get any percentages of sales, should you choose to follow my lead; just know that I've used this platform for a succession of bots and have never had any problems with it. It comes with two motors and two wheels, a single-pole, single-throw (SPST) switch, and various mounting brackets, connectors, and wire, and only costs about $35. You, of course, are free to choose another platform or to build your own; the important thing here is that you can control your platform using the motor controller board and a

Bluetooth-enabled device. Other than that, you're only limited by your imagination.

Basically you're going to want your bot platform to have two wheels, each controlled by a different motor. Those motors will be wired to the MOTOR1 and MOTOR2 outputs on the micro:bit's motor controller board. Then all you have to do is wire a power source to the POWER terminal block on the board and plug in your micro:bit. A separate power source for the micro:bit isn't necessary, because the motor controller board has an onboard power regulator that delivers the micro:bit's required 3 volts.

For that reason, however, you may want to wire a switch inline with the power supply (the Turtle platform has one onboard) so that you can easily switch off power to the board—and the micro:bit—without having to unplug wires or remove the :bit.

Your results may vary, but you hopefully will end up with something that resembles Figure 8.5.

FIGURE 8.5: Completed micro:bot

Depending on how you've wired everything, make sure that you get power to the board when you need it, and then move on to writing the program for your board.

The micro:bot Program

As you can imagine, the program for the micro:bot is going to be very similar to the Bluetooth test code you wrote earlier. However, you'll be turning pins on and off depending on the values that you send from the phone, and that will determine if the bot goes forward, backward, turns, or stops.

Back in your mbed developer account, to keep things simple, you can just clone the UART program you created earlier. To do so, right-click on the main folder and select Clone. Call the new program **buggy** and click OK.

Since you've cloned a working program, you won't need to mess with the `MicroBitConfig.h` file or any of the memory tweaks, since they're all included in your new cloned file. You will, however, need to fix the `MicroBitSamples.h` file (`buggy/source/MicroBitSamples.h`). Open that file, scroll down to the `#defines` section, and change line 40 to read

```
#define     MICROBIT_BUGGY
```

Then save the file. Now right-click on the `UART.cpp` file and clone it, calling the new program `buggy.cpp`. (Make sure you keep the file in the same place in the directory when you clone it; the final file location should be `buggy/source/buggy.cpp`.) When you're done, feel free to delete `UART.cpp` if you would like to keep your workspace clean.

Now let's edit `buggy.cpp`. The first thing to change is line 30, the `#ifdef` line. Change that to read

```
#ifdef MICROBIT_BUGGY
```

This tells the compiler to only compile this `.cpp` file if `MICROBIT_BUGGY` is defined in `MicroBitSamples.h`.

Now let's declare some pins. Delete the line

```
int connected = 0;
```

(let's assume you're always connected, since it just won't work otherwise) and add

```
int pin12, pin8, pin16, pin0 = 0;
```

These are the pins that are hardwired to the MOTOR1 and MOTOR2 terminal blocks on the motor controller board.

The `onDisconnected()` function you'll leave alone—it works just fine for our purposes. You do need to tweak the `onConnected()` function, however. Change the `while()` loop to read as follows:

```
while(1)
{
    ManagedString msg = uart->read(1);
    moveBot(msg);
}
```

and remove the line

```
ManagedString eom(":");
```

What this now does is read just one character from the serial stream, and then it calls the `moveBot()` function (which we'll show you how to write in a moment) with that one character. This way, you can type **a** on your serial terminal and the `moveBot()` function will immediately be called with a as soon as you hit Send on your device. You don't have to type a delimiter like the colon.

In case you're familiar with C and C++ and are wondering just what the heck a `ManagedString` is, it's the micro:bit's answer to C's character array (`char[]`) and C++'s `string` types. The `ManagedString` is a managed type that releases memory as needed—no garbage collection necessary—and is less prone to bugs by inexperienced programmers. It can

be constructed in several different ways, it can be manip-
ulated (concatenated, chopped up, etc.), and you can even
use operators like +, =, ==, and < and > with it.

OK, back to the program. Let's write the function that
does all the work—moveBot().

```
void moveBot(ManagedString message)
{
    ManagedString w("w");
    ManagedString space(" ");
    ManagedString a("a");
    ManagedString d("ad");
    ManagedString s("s");
    if (message == w) // start moving
    {
        pin12 = 1;
        pin16 = 1;
        drive = 1;
    }
    else
    {
        if (message == space) // all stop
        {
            pin12 = 0;
            pin16 = 0;
            drive = 0;
        }
    }
    if (drive == 1)
    {
        if (message == a) // left
        {
            pin12 = 1;
            pin16 = 0;
        }
        else
        {
            if (message == d) // right
            {
                pin12 = 0;
                pin16 = 1;
            }
            else
            {
```

```
                    if (message == s) // stop going left
or right
                    {
                        pin12 = 1;
                        pin16 = 1;
                    }
                    else
                    {
                    }
                }
            }
        }
    uBit.io.P0.setDigitalValue(pin0);
    uBit.io.P8.setDigitalValue(pin8);
    uBit.io.P12.setDigitalValue(pin12);
    uBit.io.P16.setDigitalValue(pin16);
    return;
}
```

This function is pretty self-explanatory, I think, despite the nested if/else statements. You compare the strings received, and depending on what they are, you send the bot in a specific direction. When you're finished, the final buggy.cpp should look like this:

```
#include "MicroBit.h"
#include "MicroBitSamples.h"
#include "MicroBitUARTService.h"
#ifdef MICROBIT_BUGGY
MicroBit uBit;
MicroBitUARTService *uart;
int pin0, pin8, pin12, pin16 = 0;
int connected = 0;
int drive = 0;
void moveBot(ManagedString message)
{
        ManagedString w("w");
        ManagedString space(" ");
        ManagedString a("a");
        ManagedString d("d");
        ManagedString s("s");

    if (message == w) // start moving
    {
        pin12 = 1;
```

```
        pin16 = 1;
        drive = 1;
    }
    else
    {
        if (message == space) // all stop
        {
          pin12 = 0;
          pin16 = 0;
          drive = 0;
        }
    }
    if (drive == 1)
    {
        if (message == a) // left
        {
          pin12 = 1;
          pin16 = 0;
        }
        else
        {
          if (message == d) // right
          {
            pin12 = 0;
            pin16 = 1;
          }
          else
          {
            if (message == s) // stop moving left or right
            {
              pin12 = 1;
              pin16 = 1;
            }
            else
            {
            }
          }
        }
    }
    uBit.io.P0.setDigitalValue(pin0);
    uBit.io.P8.setDigitalValue(pin8);
    uBit.io.P12.setDigitalValue(pin12);
    uBit.io.P16.setDigitalValue(pin16);
    return;
}
```

```
void onConnected(MicroBitEvent e)
{
    uBit.display.scroll("C");
    connected = 1;
    while(1)
    {
        ManagedString msg = uart->read(1);
        moveBot(msg);
    }
}
void onDisconnected(MicroBitEvent e)
{
    uBit.display.scroll("D");
    connected = 0;
}
int main()
{
    // Initialize the micro:bit runtime.
    uBit.init();
    uBit.messageBus.listen(MICROBIT_ID_BLE, MICROBIT_
BLE_EVT_CONNECTED, onConnected);
    uBit.messageBus.listen(MICROBIT_ID_BLE, MICROBIT_
BLE_EVT_DISCONNECTED, onDisconnected);
    uart = new MicroBitUARTService(*uBit.ble, 32, 32);
    uBit.display.scroll("BUGGY!");
    // If main exits, there may still be other fibers
    // running or registered event handlers etc.
    // Simply release this fiber, which will mean
    // we enter the scheduler. Worse case, we
    // then sit in the idle task forever, in a
    // power-efficient sleep.
    release_fiber();
}
#endif
```

Save the program, compile it, and flash the hex file to your micro:bit. You can download the file at https://github .com/wdonat/microbit-code/blob/master/chapter8/buggy.cpp. Once it's loaded, pair your phone with the micro:bit and bring up the Bluetooth serial terminal app you installed. After you connect to your board, type (and send) w. Your bot should shoot across the room. If you need it to stop, send a space; sending **a** and **d** should make it turn left and right, respectively. (If your bot turns right when you type **a** and left when you type **d**, it simply means that you've reversed your connections from the motors going to your motor board. Just swap the wires.) If it's turning in a circle, sending an **s** will make it stop turning and resume going forward. Again, a space will make it come to a full stop no matter what it's doing.

Now, granted, this is *not* particularly elegant code. Without writing a smartphone app, we're reduced to communicating with the micro:bit with a serial connection, sending characters one at a time. I tried to borrow from the typical video game keys, using w, a, s, and d, but I'll admit it can be tricky to get the hang of typing **w** and hitting Send and then **a** and hitting Send and then **s** and hitting Send quickly in succession in order to get the bot to turn a corner and keep going. It *can* be done, of course; my dog is now a highly trained bot chaser.

If you have experience with writing apps for your particular platform, I encourage you to write a more graphical interface for your bot. The micro:bit Blue app I mentioned earlier that is currently Android-only has a gamepad screen that can be used with a program very similar to ours to control the bot. You can find that program here: https://lancaster-university.github.io/microbit-docs/ble/event-service/. I thought it was important, however, that you first learned about the guts of Bluetooth serial communication, and of course there's the small problem that if you own an iPhone, the gamepad interface will be unavailable to you.

That concludes our introductory exploration of the micro:bit platform. If you want to get deeper into programming the board, particularly in C/C++, I highly recommend visiting https://lancaster-university.github.io/microbit-docs/. The documentation there of the micro:bit runtime is exhaustingly thorough; all you'll ever need to know about all possible classes, functions, types, and so on can be found there. If you still prefer Python and the other languages, I encourage you to stay abreast of changes; the micro:bit environment is new and is changing quickly. Either way, keep learning and building, and I look forward to seeing what you come up with!

The Story of the BBC micro:bit

So where did the micro:bit come from? Who thought of it, why, and why is it becoming so successful? And how do you go about giving away almost a *million* individual boards to children all across the United Kingdom?

To answer those questions, I corresponded with Howard Baker from the Micro:bit Education Foundation. He and Jo Claessens in their roles in BBC Learning worked together to design the first prototype, and Michael Sparks is the BBC R&D engineer who actually built it. (Howard has an interesting background: he's been a chemist, a fashion designer, a science teacher, a journalist, and a researcher.) I also chatted with Gareth Stockdale from the BBC. I was curious as to the project's inception, how it evolved, and what went into making it a success.

> **NOTE** All quoted material in this appendix came from an interview with the author conducted on May 25, 2017.

According to Howard, the idea for the micro:bit was spawned by two separate news articles that were released at about the same time in 2011 and 2012. Both articles—one from The Royal Society and one from Nesta—discussed the fact that children in the United Kingdom were graduating with almost no knowledge of coding or computer science, and the result was that the UK expected to have a "knowledge deficit" of skilled tech workers in a few years.

> NOTE If you're familiar with the origins of the Raspberry Pi, you may see a similarity here. In the case of the Pi, Eben Upton and its other creators noticed that British young adults were entering college with almost no knowledge of computer science or programming; students' idea of programming was writing a little HTML and CSS for a website.

Because Howard was then working for the BBC, he put forth the idea that the BBC should do something to "spotlight the problem, raise the profile of computer science, provide positive role models and get kids coding and interested in computer science." The organization had had previous experience—in the 1980s the BBC Micro had introduced children all over the UK to computers. Howard and his colleagues had noticed that the Maker and Hacker movements had become very popular with young people, and he wanted to build on that, to enable children to *make* something.

When it came to designing a product that could do what they wanted it to, Howard and his team considered a host of factors, including cost, complexity, and how easy it would be for kids to interact with the board. Luckily, ARM chips have been steadily decreasing in price, so they were able to make a device that was much cheaper than a Raspberry or an Arduino (both excellent boards, but costing a bit more than pocket change if it came to giving them away). As Howard puts it:

We thought of the device as a "platform"—there was the device itself—the micro:bit—but it had to be delivered with the simplest, easiest software possible and learning resources that would excite kids to get involved. It had to be very cheap so we could give it away—this gave some challenges, including its shape. It had to be very easy to program—nothing was to get in the way of a kid being given one and them getting it to do something interesting. It had to excite kids; they needed to look at it and want to get to know it and use it. However, it also needed to be easily extensible—low floor, high ceiling—once they had done something to excite them they could see the potential to build complex things with it and the device would be capable of letting them do that. It was also necessary for it to be a physical object, stripped of coverings, so a child could see how it worked and it could fit in the palm of their hand—they could touch it, work with it with their hands, build things with it. We also planned the device to be a wearable, so again it had to have a certain size and appearance. The other thing I was thinking of at the time was the Internet of Things. I was looking at what was the next revolution to affect society—preparing kids for an exciting tech revolution is a better stimulus than telling them if they didn't get their act together and learn coding they would be unemployable. I wanted the micro:bit to be an Internet of Things tool—something that would help them create the revolution, not wait to consume it.

Gareth Stockdale from the BBC adds that "there are significant and distinct educational advantages to providing a hands-on experience" when it comes to teaching kids about technology. In other words, just learning to code is one thing; learning to code on a device that then *does something* is another thing entirely. As an engineer myself, I can totally vouch for this way of thinking; I enjoy coding, but I much prefer it when my code makes something do a task that interacts with the physical world. Perhaps I'm not much more than a kid after all.

As far as partners with the project, the BBC was obviously involved straight from the beginning and has nursed the project through its current state. It has always been an extremely popular program within the organization, so although there were difficulties, there was enough in-house support to make sure it succeeded.

In December 2014, the BBC put out a call for expressions of interest for organizations wanting to be partners in the project. More than 150 companies responded, and 31 were chosen. It wasn't a problem to convince people that the micro:bit was necessary—many organizations were already aware of the looming skills shortage and were looking for ways to address it.

In fact, according to Howard, there was no shortage of companies and individuals wanting to donate cash, chips, services, software, and everything else; the hard part was coordinating everything. The BBC Micro project was thirty years ago, after all, and getting dozens of partners to work together in sync was, as you can imagine, a logistical nightmare. In the end, a total of thirty-one organizations banded together in order to make it possible for the BBC to create a board that could be given away to a million children—one for every school child in their first year of secondary school (the equivalent of seventh grade in the United States). Initial prototypes proved popular—kids loved the 25-LED matrix, the two buttons, and the fact that it was so easy to get the board to do something.

There has been support from more than the coordinating partners, of course. For example, in April 2016 the *BBC One* show aired a report showing how three students from Fallibroome Academy in Cheshire, England, were taken to the Lovell Telescope at the Jodrell Bank Observatory. There, astrophysicist Tim O'Brien allowed them to program their micro:bits to turn the 3,200-ton telescope to point at a distant pulsar (https://www.youtube.com/watch?v=NqIzufUiwN4). Not bad for a device that weighs only a few grams!

There are obviously going to be comparisons between the micro:bit and the Raspberry Pi, so I asked Howard if the Pi was part of the inspiration, and if there were things he wanted to emulate or do differently from the Pi Foundation. His response was that micro:bit team wanted to create something that had a lower barrier to entry than some of the existing products in the market. The team kept that in mind as they designed their board. They also tried to keep their target audience a bit younger than the Pi's, letting their device act as a springboard to more advanced devices like the Pi and the Arduino.

The Pi Foundation was very supportive of the project and was involved with discussions right from the beginning. The micro:bit team always kept the Pi in mind as they were building their board; they wanted a device that would be extremely easy to set up and use so that there was no entry barrier, but that would then get users interested in the next logical step—a Pi or something similar. They wanted the micro:bit to be a stepping-stone to the Raspberry Pi, and the boards seem to be working together quite well. The micro:bit has sensors that the Pi doesn't have, and the two can communicate easily over BLE. The micro:bit can then display information from the Pi on its LED matrix, so their partnership seems to be a success.

As to what the future may hold for the micro:bit . . . the organization has some grand plans. In late 2016 the BBC launched the nonprofit Micro:bit Education Foundation, with the goal of empowering children, parents, and teachers around the world to create and learn using the micro:bit. The Foundation's big objective: to get a board in the hands of 100 million people around the world. Nine partners are now involved: the BBC, Microsoft, ARM, the British Council, IET Institution of Engineering and Technology, Nominet, Amazon, Samsung, and Lancaster University. In pursuit of their goals, the team has migrated the original website, http://microbit.co.uk, to http://microbit.org, which is available in twelve different languages. They are continuing educational programs in the UK and are sponsoring classroom kits around the world, with national rollouts in progress in the UK, Iceland, and Croatia. The fact that you can now purchase the micro:bit from US retailers like SparkFun is partly thanks to the Foundation's work, and the board is now available in thirty-two countries around the world.

In closing his remarks with me, Howard stressed that the micro:bit "is the product of the very hard work of a large number of clever and wonderful people and organizations." Those organizations that have led the way on software, hardware, design, manufacture, and distribution include ARM, Barclays, the BBC, element14, Lancaster University, and many others. The project is the brainchild of a large number of extremely brilliant and talented people in BBC Learning, which leads the BBC's education strategy. That strategy includes the Make It Digital season, which was announced and championed by the BBC's Director General Tony Hall in 2015. After working with the device myself, I have to say that I'm extremely impressed with the finished product and I hope it gets a whole new generation interested in computers, programming, and engineering.

Other Programming Environments

Perhaps you're not a fan of Python (gasp!) or you're not ready to take the leap into learning a full programming language and just want to play around with your micro:bit board using some easy-to-learn IDEs.

If that's the case, you're in luck. The Micro:bit Foundation currently has three other coding environments available for the micro:bit, aside from Python: the Blocks Editor (based on Microsoft's Blocks environment), Code Kingdoms' Java-Script, and Microsoft's Touch Develop. All of these are fairly user-friendly, and in this appendix I'll walk you through setting up and using each of them.

The Blocks Editor

The JavaScript-based Blocks Editor uses a user interface that's not unlike puzzle pieces. In fact, if you've ever used the Scratch programming interface (it comes preinstalled on many Raspberry Pi distributions, for instance), you'll probably recognize it right away. In the case of the micro:bit, programming puzzle pieces vary from basic (`show number 0`) to more intricate control of the LED array (`plot x 0 y 0`) to yet more advanced functionality like reading and writing data over a serial connection or controlling and reading from the GPIO pins. It's designed the way it is to appeal to younger users, but honestly it's a pretty attractive interface for adults, too. At first glance it appears to be rather simplistic, but the interface belies its actual capabilities.

To get started with Blocks, point your browser to https:// makecode.microbit.org/. You'll be greeted by the screen you see in Figure B.1.

FIGURE B.1: The Blocks Editor screen

To create a script, you choose from the various categories in the middle, ranging from Basic to Input to Radio and Logic (see Table B.1 for a rundown of each category). Clicking on a category brings up a new screen with all of the puzzle pieces available to you (Figure B.2).

TABLE B.1: Categories and functions

CATEGORY	EXAMPLE FUNCTIONS
Basic	show leds, show icon, show string
Input	on button press, on shake, get temp
Music	play tone, start melody, change tempo
Led	plot (x,y), unplot (x,y), plot bar graph
Radio	send number, send string, receive value
Loops	repeat, while
Logic	if/then, if/then/else, and, or, not
Variables	create/set variables
Math	+, -, *, /, random numbers

FIGURE B.2: The Radio pieces

You then drag and drop the blocks onto the right portion of the screen, where they fit together as you place them. Loops and conditionals, like `while()` and `if()`, are shaped with a "hole" in them (Figure B.3). Declarative statements, meanwhile, are shaped to fit inside those holes and are processed as part of the loop that encases them (Figure B.4). The outer loop piece will resize itself to fit around the inner portion if necessary (the `show leds` piece, for instance, is a pretty hefty piece).

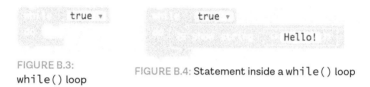

FIGURE B.3:
`while()` loop

FIGURE B.4: Statement inside a `while()` loop

Make your selection from the list on the left, and drag it into the workspace on the right. Notice that if you hover over a piece after you place it, an information dialog pops up telling you not only what the piece does, but also how it would look if you were coding it in JavaScript (which is, after all, what the end result of this interface is—a JavaScript program).

Perhaps the best way to illustrate how this works is just to jump right in. Start by clearing out any blocks currently on the workspace by right-clicking them and selecting Delete Block. If you right-click on a loop piece that encases several other pieces, you'll have the option of deleting the entire group, rather than having to do it one by one.

Once your workspace is clear, click the Basic category and drag the `on start` piece and the `show leds` piece. Drag the `show leds` piece into the `on start` piece until it clicks into position and the `on start` piece resizes to fit. (This can take some fiddling; I found that if you click on and drag the small indent at the top of the piece and drag it to the matching indent on the outer piece, it'll click right into place.)

When your two pieces are interconnected, click on some of the squares in the show leds array—they'll turn red as you click, meaning they'll be lit when you start the program. When you're satisfied with the design, leave those pieces and drag another one—let's say the on button A pressed from the Input category—onto the workspace. Finally, go back to the Basic category and drag the show string "Hello" piece onto the board. Click on "Hello" and fill in whatever string you'd like to display. Then drag that piece into the slot on the on button pressed piece. When you're done, your workspace should look like Figure B.5.

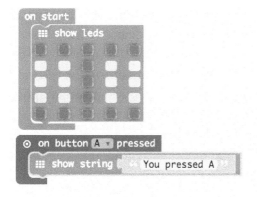

FIGURE B.5: Your first Blocks program

Now, before you download and flash your script, notice the picture of the micro:bit board on the left of your browser window. It most likely is currently displaying the design that you placed in your show leds puzzle piece. (If it's not, click the replay icon to the left of the square below the micro:bit picture.) That's because this image actually gives you a chance to preview the script you've written. If you now click the A button on the image with your mouse, the board will display the message you typed. If you're satisfied with how everything looks, give the script a new name in the text box next to the Download button. Then click Download and save the hex file to your hard drive. Drag the file onto your micro:bit, and voilà! It should work exactly as the preview image did.

Keep experimenting with all of the pieces; as you can see, you can program some pretty complex behaviors with these simple puzzle pieces.

Code Kingdoms

Upon first inspection, out of all the programming environments Code Kingdoms seems to be the most kid-oriented (when you first bring it up, it shows a Minecraft-style loading screen). However, as with Code Blocks, the simplicity of the interface belies the complexity underneath, and it has more than a few hidden gems that are not immediately apparent. To get started, point your browser to https://www.microbit .co.uk/app/#create:tomwku (or bring up https://microbit.org/ code/ and click the Code Kingdoms JavaScript icon at the bottom of the screen; see Figure B.6).

FIGURE B.6: Code Kingdoms icon

The interface is a bit more cluttered than Code Blocks (Figure B.7).

The pieces and snippets you will use to code are on the left side of the screen. The workspace where you will work is

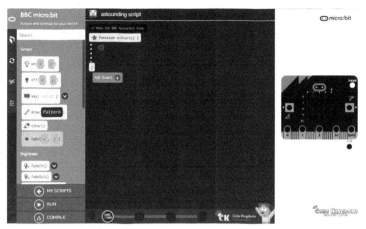

FIGURE B.7: Code Kingdoms workspace

in the middle, and the micro:bit preview application is on the right. On the far left of the screen are some additional icons; the top micro:bit logo icon is the default and shows the various actions and controls available for your script. The next icon down looks like a book and gives you access to some additional libraries: Math, Random, and Globals. Below that, the circling arrows logo gives access to control flow portions of script, such as threads, conditionals, and loops. The scissors icon below the book is where you can store snippets of code for later use, and finally the last icon in the column is a library of tutorials, ranging from a simple roll-the-dice game to a more complicated maze runner script.

Perhaps the neatest part of this interface is the slider you see across the bottom of the workspace. If you have some code loaded in the workspace, moving the slider changes the "complexity" (for lack of a better term) of the code shown. Moving the slider all the way to the left shows some simple icons relating to the code, and moving the slider all the way to the right shows the code only. Figure B.8 shows the same batch of code in all four slider positions.

FIGURE B.8: Code in four slider positions

The idea here, obviously, is that you can become acquainted with the code behind the icons, and as you get more comfortable with coding, you can leave the graphic interface behind completely.

In the meantime, however, you can experiment with the interface with the slider set in the graphic mode. It's a bit less intuitive, in my opinion, than the Code Blocks interface, but still very easy to use. When you first bring up the web page, you'll see that the function `onStart()` is preloaded. This function encapsulates everything that you program so that it'll run when the micro:bit powers on. You can create functions outside of this main function, of course, but just

Getting Started with the micro:bit

keep in mind that `onStart()` contains your main program, similar to the `while True()` statements we've used before. The Add Event button below it is where you'll enter other functions, like `roll_dice()` from our craps game, for example.

To begin, click the micro:bit logo at the top left to make sure you're in the Basic category, then click on the Draw (Pattern) icon and drag it into the workspace between the beginning and ending brackets (`{ }`) of the `onStart()` function. Then click on the word `Pattern` in parentheses and choose one of the many images available (Figure B.9). When you've chosen your image, click the green check mark to close the pop-up box.

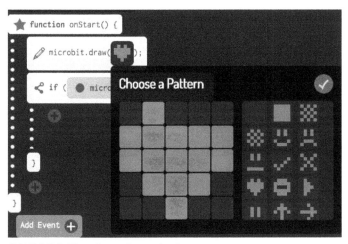

FIGURE B.9: Choosing an image to draw

Now let's clear the screen after a few seconds. In the Control Flow section (the circling arrows at the extreme left of the screen), click the `wait(milliseconds)` piece and drag it into the `onStart()` function, underneath the `microbit.draw()` line. Click the word `milliseconds` and enter **2000**. Then go back to the Basic category and drag the `clear()` button into the workspace.

You're ready for the main program script. Back in the Control Flow section, find the `while (test) { }` button and drag it into your workspace underneath the existing code but still within the `onStart()` function brackets. Click on the small arrow by the word `test` and select `true` from the pop-up window (Figure B.10).

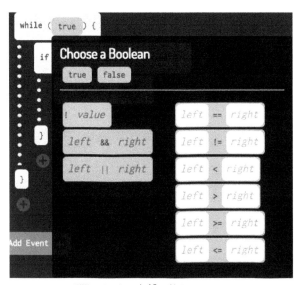

FIGURE B.10: Filling in the `while()` loop

Now click the circling arrows on the left to access the control flow blocks and find the `if (test) { }` conditional block. Drag it into the `while(True)` loop. Now go back to the Basic category on the left, scroll down to the System subcategory, and drag the `buttonAPressed` block onto the word `test` in purple on the `if()` conditional block. This is now akin to

```
if button_a.is_pressed():
```

in Python. Select another simple task to perform when the A button is pressed, such as `say(value)`. Then click on the `value` prompt and enter a string, such as `"You pressed A!"`.

If at any point you make a mistake and need to delete a block, simply click on it and drag it away from the main code. A ghostly trash can will appear at the bottom right of the workspace, and you can drop the block into the can to delete it. You can also re-input any text or conditionals you've added just by clicking on them again inside the block; the pop-up you saw in Figure B.10 will reappear and allow you to choose a different value.

That's it for a very simple program! If you're curious to see what the actual code looks like, move the slider at the bottom all the way to the right, and you should see this:

```
function onStart() {
    microbit.
draw(Pattern("01010.1111.1111.01110.00100"));
    wait(2000);
    microbit.clear();
    while (true) {
        if (microbit.buttonAPressed) {
            microbit.say("You pressed A!");
        }
    }
}
```

As you can see, it's very similar to Python in commands, but you're looking at pure JavaScript here. If you're comfortable using JavaScript, you should definitely experiment more with the window in this configuration—there's a lot you can do that isn't necessarily available via the basic kid-friendly icons.

Finally, of course, you'll want to preview the program and then load it onto your board. To preview it, click the Run button at the bottom right of the page. The interface will compile the script and—assuming everything compiles correctly—will then load it onto the simulated micro:bit on the right. You can't test things like shake or compass headings, but you can make sure that button presses and displays work right.

If you're satisfied, and if the code loaded without errors onto the simulated board, click the Compile button at the bottom to create an actual hex file. After a moment a pop-up window will let you know your script is ready to download, and you can save it and flash it onto your device.

That is a short introduction to the Code Kingdoms web-based coder. As I said, it's a complex interface. Although it definitely presents a kid-friendly appearance, my experience has been that it really isn't very intuitive, and it could definitely use some more fleshed-out tutorials. Still, you may like it, and I do like that it allows you to jump directly into coding in JavaScript.

Finally, let's take a look at Microsoft's Touch Develop.

Microsoft Touch Develop

To get started with the Touch Develop interface, point your browser to https://www.microbit.co.uk/app/#create:hrvbin. You'll be greeted by the screen you see in Figure B.11.

Like the advanced side of the Code Kingdoms page, Touch Develop seems to be aimed at introducing you to the JavaScript foundations behind the scripts you will be writing. The workspace, complete with a starter script, takes up the majority of the page. At the bottom is where you'll find the snippets, functions, and variables that you can put into your script, and on the right is the simulated micro:bit board where you can test your code before uploading it to your actual device.

There are two main ways to interact with your code: you can choose snippets from the bottom, which populates the script bit by bit, or you can actually click in the body of the script and type your code. When you start typing,

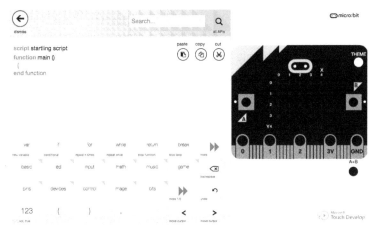

FIGURE B.11: The Microsoft Touch Develop interface

however, you're not left out in the cold; as soon as you type a letter, the interface gives you up to ten different suggestions as to what you might want to put there (Figure B.12 illustrates what happens when you type **t**, for example).

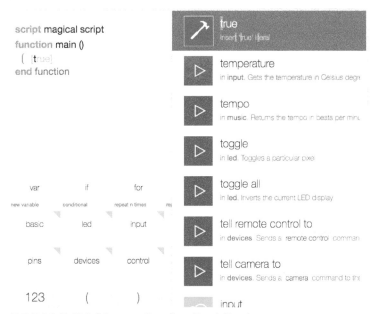

FIGURE B.12: Helpful suggestions from Touch Develop

If you just want to play with the available functions, using the buttons on the bottom is a good way to become acquainted with this interface. Starting from the welcome screen, make sure your cursor is blinking on the line directly below the `function main()` line. Click the `led` button at the bottom. The code will show the word `led`, along with the helpful prompt `"we have a led here; did you want to do something with it?"`.

Ignoring the smart-aleck interface for a moment, if you look back at the bottom of the screen, you'll see that it's now popu-lated with things you can *do* with that LED: `plot`, `unplot`, `point`, `brightness`, `toggle`, and so on. Clicking on `plot`, for example, continues to fill in the line of code with `led → plot(0, 0)`, with the cursor blinking at the first `0`. A horizontal bar at the bottom of the workspace now describes the function you've chosen, how to call it (its parameters), and what it does.

If you change your mind about a line of code, you can click on the line you don't like and three icons will pop up, letting you paste, copy, or cut the code. Cut removes the line and lets you start over.

Some of the buttons at the bottom are self-explanatory, like declaring a variable or adding a `for` loop. Others, how-ever, can take you deeper down the rabbit hole than you may care to go, at least until you're more comfortable with the board and the interface. For example, try clicking the Game button.

Once again the interface asks if you're going to do any-thing with that nice `game` object, but now the bottom of the page is filled with functions like `create sprite` and `add score`. Clicking the `create sprite` button adds the line `game → cre-ate sprite(2, 2)` to your code, which creates a new LED sprite pointing to the right. `start countdown`, on the other hand, adds the line `game → start countdown(10000)`, allowing you to start a game countdown timer. Obviously you can use the Touch Develop interface to program an entire game from scratch, using the predeveloped libraries that Microsoft has made available to you.

If you click the magnifying glass icon located toward the top of the screen labeled "all APIs," you'll see a list of the various categories of events and functions that you can play with (Figure B.13) and a short description of what each one does. There's a little bit of everything here, from basic functions like displaying strings, to controlling paired devices, to performing bitwise arithmetic on 32-bit integers. Clicking on any one of them brings up its associated variables and functions at the bottom of the screen.

FIGURE B.13: Additional libraries of functions

Let's create a basic script and upload it to our device. From the starting page, click on the empty line below `function main()` to bring up the row of functions at the bottom. Click the `while` button, which will automatically fill in the loop with the following:

```
while true do
    (add code here
    (basic → pause(20)
end while
```

Clicking on the `add code here` line will bring up the list of available functions again; choose the `basic` button and then `show string`. Type **Hello, world!** in the text box, and the line will autocomplete with `basic → show string("Hello, world!", 150)`, which, according to the informational block, means that it will display the text, one character at a time, shifting by one column each 150 milliseconds.

That's a good basic, introductory script. Click the Dismiss arrow at the top left of the screen, and you'll see some new icons: My Scripts, Run Main, Compile, and Undo. The My Scripts button takes you to a page of scripts that you've worked on, Run Main runs your program on the simulated micro:bit board, Compile compiles the script so you can flash your board, and Undo reverses anything you may have done. Don't worry about running this particular script on the simulator—it's a very basic script, after all. Rather, click the Compile button and save the hex file when you see the message that it's ready to download. Drag that downloaded file onto your micro:bit, and voilà! You've just used the interface! Your board should slowly scroll the text "Hello, world!"

There's a ton more material here to go through, but this isn't a book about these interfaces. All three listed here have some very good tutorials available; if you're interested in learning more about them I highly recommend working through the online lessons. Personally, I prefer the Python and C++ environments (via mu and mbed), but these environments are just as effective.

Index